Video Compression
Demystified

The more predicable the event the less information
is carried by the event

0 775 711 65 97

McGraw-Hill Video and Audio Books

Video Compression Demystified

Peter Symes

McGraw-Hill
New York   San Francisco   Washington, D.C.   Auckland   Bogotá
Caracas   Lisbon   London   Madrid   Mexico City   Milan
Montreal   New Delhi   San Juan   Singapore
Sydney   Tokyo   Toronto

Library of Congress Cataloging-in-Publication Data

Symes, Peter (Peter D.)
 Video compression demystified / Peter Symes.
   p.   cm.
 Includes bibliographical references and index.
 ISBN 0-07-136324-6
 1. Video compression. 2. MPEG (Video coding standard)
 3. JPEG (Image coding standard) I. Title.
 TK6680.5.S95   2001
 621.388—dc21   00-066425

McGraw-Hill

A Division of The McGraw·Hill Companies

 2 3 4 5 6 7 8 9 0   DOC/DOC   0 6 5 4 3 2 1

P/N 136382-3
PART OF
ISBN 0-07-136324-6

*The sponsoring editor for this book was Stephen S. Chapman and
the production supervisor was Sherri Souffrance. It was set in
Vendome by Patricia Wallenburg.*

Printed and bound by R. R. Donnelley & Sons Company.

Portions of this book were previously published in
Video Compression, © 1998.

McGraw-Hill books are available at special quantity discounts to use
as premiums and sales promotions, or for use in corporate training
programs. For more information, please write to the Director of Special
Sales, Professional Publishing, McGraw-Hill, Two Penn Plaza, New York,
NY 10121-2298. Or contact your local bookstore.

 This book is printed on recycled, acid-free paper containing a
minimum of 50% recycled, de-inked fiber.

CONTENTS

Contents

PREFACE

This book is a revision and extension of my book *Video Compression* published by McGraw-Hill in 1988. It includes some corrections, additional explanations, and new chapters on compression techniques and standards that have come to the fore in the last two and a half years. The new book also includes a CD-ROM with software tools and sample material that will allow the reader to experiment with some of the systems discussed.

I have received a variety of comments on the first book. Some have been critical because the book does not offer implementation instructions or much detail of the syntax of Standards like MPEG. I must emphasize that this was not the objective of the book. There are excellent texts (some referenced in the bibliography) that cover these subjects and, as supplements to the Standards documents themselves, are recommended reading for anyone designing compression systems. At the other extreme it has been really gratifying to receive comments from compression system designers who found that my book helped them to understand the fundamentals of the systems they were implementing.

Compression technology has come a long way since the first book, and some of the developments are covered in this volume. More importantly, the impact of compression technology has become really substantial. Digital cable and satellite systems have achieved widespread acceptance; terrestrial digital television is having birthing difficulties in North America, but is very successful elsewhere. DVD players and personal video recorders are both impacting the way we watch television. But in reality, all these pale in comparison to the impact of Napster, MP3.com, and the other distributors of compressed music files. I make no comments on the rights and wrongs of the issues, not do I have any proposals to reform the copyright laws. I do believe that these organizations have made us realize that the world is changing. We will have to learn to think in terms of digital assets rather than pressed vinyl or videotape.

Perhaps it is best expressed by the T-shirt of a Microsoft employee seen recently, that announced "You do not have the right to remain analog!"

Peter Symes
November 2000

ACKNOWLEDGMENTS

My understanding of many of the topics in this book is due to the clarity of explanation of Dr. Majid Rabbani of Eastman Kodak at an excellent course organized by Portland State University. His explanation of "DCT as Axis Rotation" certainly helped me become more comfortable with DCT, and I am grateful for his permission to use it here.

There are a number of references in this text to Charles Poynton's book, but Charles' impact is much greater than that. I owe to him much of my understanding of the fundamentals of vision, imaging, and digital systems. Others, including Bruce Penney, Kerns Powers, Fred Remley, Charles Rhodes, and Larry Thorpe are also members of this special group.

The analysis of concatenation of identical compression systems included in Chapter 19 is the work of Jim Wilkinson, Chief Research Scientist of Sony Broadcast & Professional, Europe; his permission to present it in this book is gratefully acknowledged.

The section "Practicing the Art of MPEG" in Chapter 10 is based heavily on a presentation by Dr. Michael Isnardi, and my thanks are due to him and to the Sarnoff Corporation for permission to include this material.

My thanks also to Jim Blecksmith of Grass Valley Group, who constructed some of the DCT examples shown at the end of Chapter 5 and allowed me to use them here.

Less formally, most of my knowledge of this subject has been gleaned from my colleagues in the industry. I hesitate to make a list because inevitably someone of significance will be omitted; however some people must be mentioned.

Mike Isnardi was not the only person at Sarnoff Corporation to help in my education. Glenn Reitmeier, Norm Hurst, Terry Smith, and Joel Zdepski (now with Thomson Consumer Electronics) are among the many who generously shared their knowledge.

SMPTE and ATSC committees are always educational and I have learned from almost everyone who participates. My particular thanks to Stanley Baron, Katie Cornog, David Fibush, Robert Hopkins, Bernard Lechner, Steven Lyman, Bill Miller, Johann Safar, Craig

Todd, and Merrill Weiss—and my sincere apologies to those I have forgotten to mention.

It was Leonard Dole who suggested I write the first book, and Jerry Whitaker who supplied important encouragement and advice when it was most needed. My thanks to them, and to my colleagues at Grass Valley Group for their encouragement and lack of obvious skepticism!

My friend and long-term colleague at Grass Valley Group, Wayne McLachlan, deserves a special mention. We have taught video compression together at UCLA and elsewhere, and Wayne's insight and thoroughness have helped me understand better many of the topics in this book, and to clarify the explanations. He also spotted some of the most embarrassing mistakes in the first book!

Despite all this expertise at my disposal, I know there will still be mistakes in this book. I bear sole responsibility for these.

The staff at McGraw-Hill were ever helpful; my particular thanks to Steve Chapman and Petra Captein. Patty Wallenburg of TypeWriting did the production of this book, and was constructive and supremely patient.

I think that the CD-ROM included with this book will add a great deal to its value. Most of the software is available elsewhere, but I hope that collecting it together and packaging with some source material will make it much easier to experiment. My thanks to all the people and companies who allowed me to distribute their work in this way. They are too numerous to list here, but all are referenced on the CD.

I have to thank my family, and particularly my wife Stephanie, for tolerating the persistent piles of books and paper, not to mention my moods on occasion. Most important though, my heartfelt thanks for their support and encouragement.

Peter Symes

Video Compression
Demystified

What Is Compression?

Introduction

Compression is the science of reducing the amount of data used to convey information. In this book we will examine a wide range of techniques—some simple, some very complex—and ways in which these techniques may be applied to video.

Compression relies on the fact that information, by its very nature, is not random but exhibits order and patterning. If this order and patterning can be extracted, the essence of the information can often be represented and transmitted using less data than would be needed for the original. We can then reconstruct the original, or a close approximation of it, at the receiving point.

There are several families of compression techniques, fundamentally different in their approach to the problem. These techniques are in fact so different that they can often be used sequentially to good advantage. Sophisticated compression systems use one or more techniques from each family to achieve the greatest possible reduction in data.

There are arguments about the semantics of compression. Some authorities use the term *compression* only when there is some resultant impairment, and they prefer terms like *bit rate reduction* for techniques that involve no loss. In this book *compression* is used to cover all techniques that reduce data, whatever the nature of the process and whether or not errors are introduced.

Before examining compression techniques, let us look at the general issue of information and the data used to represent it.

Information and Data

A major concern in today's world is the handling of information. The information may be of many types—written text, the spoken word, music, still pictures, and moving pictures are just a few examples. Whatever the type of information, we can represent it by electrical signals or data, and transmit it or store it.

The more complex the information, the more data is needed to represent it. Plain text can be represented by 8 bits/character, or about 20 kbits for a page. CD-quality music requires nearly 1,500 kbits for each second, and full-motion 525-line video (as used in North American television systems) needs about 200 Mbits (200,000 kbits) for each second.

When we need to transmit a given amount of data, we use a certain bandwidth (measured in bits per second) for the time necessary to transmit all the bits. In the real world, there is always some restriction on the available bandwidth. For example, if the transmission must be over regular telephone lines using modems, it is difficult to achieve more than 33.6 kbits/s (kilobits per second). This is fast compared to the 300-bits/s modems of a few years ago, but even at 33.6 kbits/s it would take some two hours to transmit a single second of high quality video.

Sometimes it is a practical restriction, like the use of a modem, that sets the requirements for compression. We could send high-quality uncompressed audio data over a modem, but each second's worth of audio would take about 50 seconds to transmit. In other words, we would have to receive the data gradually, store it away, then play the resulting file at the correct rate to hear the sound. However, we may want to transmit sound so that we can listen to it as it arrives in "real time." We could choose to reduce the quality we are trying to send, but even telephone-quality speech needs a data rate substantially higher than our high-speed modem can provide. If we want to send real-time audio over a modem link, we have to compress the data.

Another example of this type of requirement is the digital television system developed for North America. When the study was started in 1977, few thought it would be possible to transmit high-definition television (characterized by the 1125-line system) over a single 6-MHz channel as used for today's 525-line television. When digital techniques were considered, the problem looked even worse, because the high-definition signal represented over 1 Gbits/s of data. In the event, two teams of engineers combined to produce the answer: the transmission engineers designed a system to deliver nearly 20 Mbits/s reliably over a 6-MHz channel, and the compression engineers achieved data reduction of about 60 to 1, while maintaining excellent picture quality.

At other times the motive for compression is financial. Bandwidth costs money, and so compression reduces the cost of transmission. A satellite link with a data bandwidth of 40 Mbits/s will earn more money if it can carry ten television channels instead of just one. I may want to download a 6-MB file from a computer online service; if a compressed version is available that is only 1.2 MB I will save 80 percent of the online charges.

Transmission and storage of information might seem to be very different problems but, in terms of quantity of data, the economics are the same. Just as there is a cost for each bit transmitted, there is a cost for each bit stored in memory, on disk, or on tape. If we can use less data, both transmission and storage will be cheaper. There are some differences in how compressed data is best structured, and these will be mentioned in Chapter 10. Otherwise, everything we say about compression applies equally to transmission or storage.

The motives for wanting to reduce the amount of data should be clear. Now we will examine some of the techniques.

Information Reduction

One important step in compression technology is easy to ignore because it lacks the glamor of some of the sophisticated techniques. It is extremely important to ensure that the signal or data stream we want to compress represents the information we want to transmit, and *only* the information we want to transmit. If the data stream contains surplus information of any nature, this information will take bits to transmit and result in fewer bits being available for the information we do need. Surplus information is *irrelevant* because the intended recipient can make no use of it.

Surplus information can take many forms. It can be information in the original signal or data stream that exceeds the capabilities of the receiving device. For example, there is no point in transmitting more resolution than the receiving device can use, and no point in transmitting color information if the receiver is known to be monochrome.

Another form of surplus information is *artifacts*—features or elements of the input that are not truly part of the information. *Noise* is the most obvious example; almost anything derived from a physical transducer has some component of noise. Noise is by nature random or nearly so and, as we shall see, this makes it essentially incompressible. In video, noise may come from many sources including the transducer, analog processing, and film grain. When the signal has been converted to digital form, it may have additional *quantization noise,* and digital processing may add noise through truncation.

Many other types of artifacts exist, ranging from filter ringing to film scratches. Some may seem trivial, but in the field of compression they can be very important. Compression relies on order and self-consistency in a signal. An artifact—anything that does not "belong"— tends to require a disproportionately high number of bits to transmit, just because it destroys this order and self-consistency.

This book concentrates on the techniques of compression. Preprocessing of video to remove irrelevant information and artifacts is a very complex subject in itself. A preprocessing unit developed by Snell & Wilcox includes no less than seven different filters!

Lossless Compression

Some compression techniques are truly lossless. In other words, when we have compressed some data, we can reverse the process (*decompress*) and get the same data we started with, exactly and precisely. Lossless compression works by removing *redundant* information—information that, if removed, can be recreated from the remaining data.

In one sense this is the ideal of compression; there is no cost to using it other than the cost of the compression and decompression processes. Unfortunately, lossless compression suffers from two significant disadvantages. Typically, lossless compression offers only relatively small compression ratios, so used alone it often does not meet economic needs. Also, the compression ratio is very dependent on the input data. Used alone, lossless compression cannot guarantee a constant output data rate such as might be required for a transmission channel.

For some applications, lossless is the only type of compression that can be used. If we wish to transmit a binary file such as a computer program, receiving an approximation to the original is of no use whatsoever. The program will run only if replicated exactly. Fortunately, many computer files have a high degree of order and patterning, so lossless compression techniques do yield useful results. Compression programs like the well-known PKZIP use a combination of lossless techniques and are particularly effective on graphics image files.

One advantage of lossless compression is that it can be applied to any data stream. Most video compression schemes use lossy techniques to achieve large degrees of compression. Lossless techniques are then applied to the resulting data stream to reduce the data rate even further. Two lossless techniques are in general use.

Run-Length Encoding

Many data streams contain long runs of a value. For example, a graphics file will typically contain values for each pixel, ordered line by line. A red object will likely be represented by several long runs of the value "red." With appropriate coding it can use much less data to say "red 23 times" instead of "R R R R R R R R R R R R R R R R R R R R R R R." This is run-length encoding—we code the length of the "run" of a certain value. In video compression, various tricks can be played to make the data suitable for run-length encoding.

Entropy Encoding

Entropy encoding is usually the last step in compression. Where run-length encoding relies on adjacent values being the same, entropy encoding looks at the overall frequency of specific values, wherever they are located. The implementation of entropy encoding is quite complex, but the principle is easy to understand if we define our terms carefully.

If the data is represented by bytes (groups of 8 bits), each byte can have any *value* between 0 and 255_{10}. When we come to transmit this data we will choose a *symbol* to represent each possible value, and we transmit a sequence of *symbols* until all the *values* have been sent.

The simplest way to code the data is to send the bytes themselves; in other words, each symbol is the same as the value it represents. This is so obvious that we tend not to think of anything else, and make no distinction between the value and the symbol. But it does not have to be like this. So long as both the transmitting and receiving points know and apply the same set of rules, we can use any set of 256 unique symbols to represent the 256 possible values. This is not very helpful until we add another factor: provided we have a set of rules that lets us know when one symbol ends and another begins, *not all the symbols need to be the same length.*

This is the essence of entropy encoding. It compresses the data by using short symbols to represent values that occur frequently, and longer symbols to represent the values that occur less frequently. Unless the input data is very close to random in its distribution of values, this technique means that total data used for the symbols will be less than the total of the original values.

Optimization of entropy encoding is complex, but the basic idea is neither difficult nor new. Let us compare two ways of coding the alphabet. Anyone who has spent much time with computers has come across the ASCII encoding system. In ASCII we represent any letter by a particular set of 8 bits, or 1 byte. It does not matter what *value* (letter) we choose, the *symbol* is 8 bits long.1 A much earlier method of encoding the alphabet was invented by Samuel Morse in 1838. The Morse code is a simple but useful example of an entropy coding system.

The Morse code uses combinations of dots and dashes as symbols to represent letters. A dash is 3 times as long as a dot; each dot or dash within a symbol is separated by one dot length; each symbol is separated by one dash length. Morse designed his code to transmit English-language text, and to maximize the speed of transmission he looked at the frequency with which the various letters of the alphabet appear in typical text. The letter *e* is the most common letter in English, so Morse assigned the shortest possible symbol—one dot (actually four dots in length because of the intersymbol space). The next most common letter *t* received the next shortest symbol, one dash. At the other extreme, an infrequent letter like *y* receives three dashes and a dot, a symbol four times as long as that used for *e*, as shown in Figure 1-1.

Figure 1-1
Morse Code provides a simple example of variable length encoding.

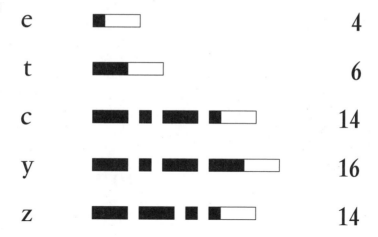

e		4
t		6
c		14
y		16
z		14

The Morse code also illustrates one of the problems of entropy encoding. For transmission to be successful, obviously both the transmitter and the receiver must use the same mapping of symbols to values. The relationship between values and symbols is known as the *code*

---

[1] Of course, ASCII encodes numbers as well as letters; it encodes uppercase and lowercase letters, and it encodes punctuation and other symbols. If we need to encode the letters only, without regard to case, we do not need 8 bits because there are not 256 letters. The old telex coding using 5-bit symbols would be sufficient. However, the comparison of fixed and variable symbol lengths is still valid.

table. But the most efficient mapping can be determined only when the frequency of values is known. The code table for Morse is derived from the frequency of letters in English text. If we wanted to send Polish-language text, where the letters *c, y,* and *z* are used more frequently than in English, this code table would not be very efficient because those letters have long symbols. If we wanted to transmit a few lines of Polish in the middle of many pages of English, this is probably not a major concern. However, if we had to transmit a very large volume of Polish, we might want to change the code table. We can do this if we have established rules at the transmitter and receiver to allow for different code tables, but we would need to evaluate this decision carefully. If we decide to change code tables, we have to transmit the new table, and that means transmitting a considerable amount of additional data!

There is no perfect solution to this, and the best compromise depends on the application. Some coding schemes use only one code table; this may not always be the most efficient table, but the coding system does not carry any of the overhead necessary to change tables. Other systems have a small number of fixed code tables, known to both transmitter and receiver. The encoder can examine the data to be transmitted, and select the most efficient code table. The only overhead is the provision of a message that tells the receiver "Use table number 3." Still other schemes allow for the transmission of custom code tables whenever the characteristics of the data change substantially.

Lossy Compression

Lossless compression would be an ideal answer, except that it rarely provides large degrees of compression, and it cannot be used alone to guarantee a fixed bit rate. Lossless compression is an important part of our tool kit, but on its own it does not provide a solution to many practical problems. If we want to put digital audio over a modem link, or high-definition television through a 6-MHz channel, we have to accept that the compression process will result in some loss; what we get out will not be exactly the same as what we put in. This is the field of *lossy compression.* Ideally, lossy compression, like information reduction, removes *irrelevant* information. Some information is truly irrelevant in that the intended recipient cannot perceive that it is missing. In most cases we also look for information that is close to irrelevant, where the quality loss is small compared to the data savings.

The objective of lossy compression is simple. We want to get maximum benefit (compression ratio, or bit rate reduction) at minimum cost (the loss in quality). However, the realization of this objective is not simple. As we will see when we look in detail at lossy compression techniques, a very large number of parameters have to be chosen for any given implementation, and many of these parameters must be varied according to the dynamic characteristics of the data. Even with a simple measure of the resulting quality (or lack of it), optimizing such a large number of variables is a complex task. Because these schemes rely on a knowledge of how the information will be perceived by the recipient, they are known as *perceptual coding systems.*

Unfortunately there is as yet no simple measure that can be applied to the quality of a compression system, particularly for video. The only measure that really matters is the subjective effect as perceived by a viewer, and this is an extremely complex function. If compression loss were confined to one characteristic of a picture, it would be relatively simple to derive a way of measuring this characteristic objectively. We could then perform careful subjective tests on a representative sample of viewers and arrive at a calibrated relationship between this characteristic and subjective quality. Unfortunately, compression loss results in not one but many changes to the characteristics of the picture; each of these has a complex, nonlinear relationship to subjective quality, and they interact with each other! (As an indication of the difficulty of quantifying compression loss in images: workers in the field consistently observe that the perceived quality of a picture improves as the quality of the accompanying sound improves.)

Compression loss in images involves two main components: things that should be in the picture, but are lost, and things that are added to the picture (artifacts) that should not be there. Areas where we would expect loss include spatial resolution (luminance and color, probably to different degrees) and shadow and highlight detail. Where extreme compression is needed for applications like video conferencing, or the Internet, temporal resolution would also be sacrificed. Artifacts that might be added include blocking, "mosquito noise" in edges, quantization noise, stepping of gray scales, patterning, ringing, and the like. As a further complication, in the presence of loss—say of resolution—the addition of small amounts of artifact such as noise or ringing may actually improve the subjective quality of the picture.

Even within "standardized" compression schemes such as MPEG, the tuning of a system to yield the maximum quality per bit is a black art involving models of the human psychovisual system that are closely guarded commercial secrets. Evaluation of the effect of parameter changes is very complex for the reasons discussed above.

There may be light at the end of the tunnel. In the fall of 1997 Tektronix released a product called the *signal quality analyzer*. This device uses a technology known as *just noticeable difference* (JND) developed jointly by Tektronix and Sarnoff Corporation. It compares an output sequence from a compression system with the known input sequence, and weights each error according to its type, severity, location, and the masking effect of other nearby picture elements. The intent is to model the behavior of the human eye-brain combination, and initial tests show an encouraging degree of correlation between the measurements made by the system and subjective testing with observers.

Image Compression Standards

Any practical compression system requires the use of a number of compression tools. These tools are combined to provide a cost-effective and optimized solution for the particular problem at hand. In proprietary systems where compression is used only internally, the designer has a free hand in selecting what tools will be used, and how. In a professional videotape machine such as the Sony Digital BetaCam®, the compressed data is used only within the machine. All connections to the outside world are made as standard uncompressed video, so that there is no need for anyone outside the manufacturing organization to know the details of the compression scheme.

This approach works as long as the compressed data stays within one manufacturer's system. We may wish to send compressed data over a network or a broadcast channel, or place it on a CD, and have it usable by a different manufacturer's equipment at the other end. In this case, everyone needs to know how to decompress the data, even if they do not know exactly how the compression was achieved. In other words, we need one or more standardized compression systems.

In the field of electronic imagery, standardization work has been undertaken by the International Organization for Standardization (ISO) in cooperation with the International Telecommunications Union and the International Electrotechnical Commission (IEC). The

Joint Photographic Experts' Group and the Motion Picture Experts' Group produced the well-known JPEG, MPEG-1, and MPEG-2 standards that form the basis of most image compression today. In 1996 the National Academy of Television Arts and Sciences recognized the importance of this work by the award of an Emmy statue to ISO and IEC.

The JPEG and MPEG standards, and the continuing work of the MPEG committee, are discussed in later chapters.

Symmetric and Asymmetric Systems

Compression systems are, in general, complex and quite expensive. One characteristic important in the choice of a system is the degree of *symmetry* or *asymmetry*. This is best illustrated by examining two different scenarios.

Consider the application of videoconferencing, a subject of great interest to compression specialists. To gain popularity for other than special events, videoconferencing must be inexpensive and convenient. To be inexpensive it must use a minimum of bandwidth, and the transmitting and receiving equipment must be affordable. To achieve low bandwidth we need compression, and every conference station needs the ability to compress and decompress. This is a symmetrical situation—there is no benefit in designing a system where compression is more or less complex and expensive than decompression.

In contrast, some applications such as television broadcasting are characterized by a single compressor feeding a very large number of decompressors. CD-ROMs and digital video disks (DVDs) have similar economics; one compression system is used to produce the master, but millions of players decompress the data. These are asymmetric models. If we can reduce the data rate, or simplify the decompressor, by additional complexity in the compressor, the system as a whole benefits economically.

Baseline JPEG is very close to a symmetric system. Encoder and decoder complexities are very similar. In contrast, MPEG, with its use of motion compensation, is an asymmetric system. The bit rate is reduced with relatively little increase in decoder complexity, but at the expense of much greater encoder complexity and cost.

Why Do I Care?

It is tempting for the television engineer to conclude that compression is a subject for experts, and that it takes place within a black box. In many ways this is valid; an engineer designing or operating a television facility does not need to be able to design a television camera, or a routing switcher, and the same is true for compression systems. However, good engineers understand in principle what is happening in each of the major components of a system, even if they could not design that component.

Compression will exist at some point in virtually all television systems. The performance of the compression is affected, sometimes dramatically, by characteristics of the video signal. A signal that has been compressed carries a signature, often invisible, that can radically alter the behavior of downstream operations.

I believe that video engineers must understand the processes that are being applied to their images. Only in this way can they protect quality and diagnose problems. Anyway, compression is a fun subject!

Upcoming Topics

In Chapter 2 we look at many of the fundamentals of the sampled images that form the actual input to any compression system. Many of the topics covered are not specific to compression, but all are fundamental to its operation. Most engineers will be familiar with this material, but I encourage you to read the chapter, anyway. These ideas are absolutely essential to the operation of compression systems, and many so-called experts go wrong because they fail to remember or apply the fundamentals discussed here.

Chapters 3 to 6 cover the basic tools of compression systems for static images, and Chapter 7 brings these tools together in a description of the JPEG system. Chapter 8 introduces the special characteristics of moving images and the technology of motion compensation. Having dealt with that, we can then move on, in Chapters 9 and 10, to MPEG-1 and MPEG-2.

Video Compression Demystified concentrates largely on JPEG and MPEG because these are the most important systems for the video engineer today. Although there are more modern techniques, the large number of MPEG-2 decoders deployed in digital TV receivers (including satellite receivers) and in DVD players is sufficient to ensure that this standard will be important for many years. However, the techniques used in these standards are not the only ones. Chapter 11 describes the DV compression system, which uses the same technology, but in a quite different way. Later chapters explore the principles and application of other technologies, including wavelets and fractals, and the fundamentals of audio compression.

No book on compression would be complete without a look at streaming video and audio on the Internet, and this leads to a description of the contents of the CD-ROM included with this book. Here you will have the opportunity to explore some of the concepts directly through demonstration programs and sample images and sequences.

The book is organized in a progressive manner—each chapter assumes knowledge of at least the basic elements of previous chapters—so it is a good idea to work through the book in the correct sequence. However, each chapter covers the basics first, and then goes into greater detail. The fundamentals are important, and much of the detail is provided for example, and because I find it fascinating. If the detail gets too much, move on to the next chapter—I had fun writing this book, and I want you to enjoy reading it!

An Introduction to Images

Introduction

In Chapter 1 we looked at the three main processes likely to form part of a compression system: information reduction, lossy compression, and lossless compression. Lossless compression is purely a mathematical exercise; it can be applied to any data stream. We need some understanding of the statistics of the data stream, which are dependent on the source, but otherwise no knowledge is needed of what the data stream represents. Information reduction and lossy compression, however, are both processes that are specific to the data type. They require a detailed understanding of the characteristics of the data, how they are generated, and how they will be used. To build a successful compression system for video we must understand the characteristics of images, and how images are converted to data. Also, we need to know the characteristics of the display and the viewing conditions, and to understand the behavior of the human psychovisual system.

In this chapter we discuss different types of images and their applications. We then examine the sampling mechanism by which an image is converted to data, and the limitations of this process.

Video or Computer Graphics?

Video or computer graphics? This is not just an idle question, nor is it an issue of purism or quality. Video (in the sense used by television engineers) and computer graphics are very different. They have different purposes, different processes, different characteristics, and they require totally different compression techniques.[1] A computer system can display video, and a television system can display graphics, but typically neither will do a very good job.

This book is about compressing video, not computer, graphics, although some of the techniques discussed in Chapter 3 are certainly applicable to graphics. Contrasting the two can help us understand

---

[1] There are, of course, areas of overlap. In the following paragraphs I discuss "traditional" computer graphics targeted mainly at business systems. However, the most sophisticated computer graphics systems can render photo-realistic imagery. To the extent that this is successful (and modern systems can be very successful) the resultant images have the characteristics of photographic images or video, and are quite suitable for compression using the techniques described in this book.

why video behaves as it does, why we use certain compression techniques, and how to make the compression system's job as easy as possible. Let's look at what each is trying to achieve.

Video, in the television sense, is about the display of photographic-style images, generally of real-world scenes and objects. We try to make edges appear natural, and expect this to apply whether the edges are horizontal, vertical, or anywhere in between. Video is essentially a continuous-tone world; we expect brightness and color gradations to be smooth and bandwidth-limited, and to stay that way.

The computer world, on the other hand, is chiefly concerned with precision and the display of idealized objects or primitives like lines and letters, rather than real-world scenes. It is important to maximize the amount of information that can be displayed on the screen, and to clearly delineate edges. Usually, this is more important than maintaining a "natural look."

To see the importance of these differences, and to understand how to compress video, we need to review sampling theory.

Sampling and Quantization

Sampling is the process of examining the value of a continuous function at regular intervals. We might measure the voltage of an analog waveform every microsecond, or measure the brightness of a black and white photograph every one-hundredth of an inch, horizontally and vertically. If we define the precision we need and then do the sampling job properly, we can subsequently take those samples and reconstruct the original *to that defined degree of precision*. For an image, we will sample one or more parameters representing luminance or color functions. Irrespective of the function being sampled, we will refer to sampling the *intensity* of the image.

Quantization is the process of limiting the value of a function at any sample to one of a predetermined number of permissible values, so that we can represent it by a finite number of bits in a digital word. If we are measuring temperature outside a house, we might be interested in a range of 0 to 120°F. If we want to record the temperature to a precision of 0.01°, we must be able to represent 12,001 possible values, and this requires 14 bits to represent each sample. If, on the other hand, an accuracy of ±5° is sufficient, we need to represent only 13 distinct values (0, 10, 20... 120°), and require only 4 bits to represent each sample. In the field of image processing for video, the quantization precision is most commonly either 8 bits (256 possible levels) or 10 bits (1,024 possible levels).

Note that sampling and quantization are separate and distinct processes, even though they are often performed within the same functional block, the analog-to-digital converter. Theoretically, the steps can be performed in either order, but it is usually convenient to sample, then quantize. Each step has its own effect on the precision of the analog-to-digital conversion process; each helps determine the amount of data we get from that process, and each affects the accuracy with which we can reconstruct the original.

We must look at sampling in some detail, but before we leave quantization let's tag a couple of points that will be important later on:

1. Quantization adds noise, in that the quantized signal contains deviations from the original. This noise is called, surprisingly enough, *quantization noise*. The lower the number of bits, the larger the quantization steps, and the higher the quantization noise.

2. Using a higher number of bits generates more data. However, the quantization noise is lower, so the data is *not necessarily* more difficult to compress. If the extra bits mean that we are sampling noise in the original signal, nothing is gained. If the original signal is relatively noise free, so that additional bits indicate a closer approximation to the original *information* (less quantization noise), we may have an easier job when we come to compress.

Sampling Rate and Aliasing

When we sample a signal, we look at its value only at specific points in time or space. Between the samples we know nothing about what has happened. Provided we sample often enough, we will generate a series of samples that allow us to reconstruct the original information. However, if we do not sample often enough, not only can we lose information, we can generate false information. This is a very important concept, so I will illustrate the point. Let's look at some simple examples.

One-Dimensional Sampling

A simple case of one-dimensional sampling is the reporting (on a regular basis) of the trading value of a stock, or the value of a stock index such as the Dow Jones Industrials. If we are not stock traders, most likely the value we hear for the Dow is the closing price each day; in other words, the value of the index at 4 P.M. Eastern Time on that day. If we gather this information each day, we will have data that represents the behavior of the index on a daily basis. However, if we listen to the financial news during any day, we will hear various values for the Dow reported at different times. Sometimes the Dow will swing up and down several times during the trading day. Our daily closing value is a perfectly valid way of tracking the medium-to long-term behavior of the index, but gives no information whatsoever about the short-term variations that may (or may not) have occurred during a day.

To go a step further we need to make assumptions about how a system behaves. If I knew how to do that with stocks I would be rich already, so let's look at a contrived system whose behavior we can understand. We will attach a light or some form of marker to the rim of a wheel, and measure the height of the marker above or below an axis through the center of the wheel (see Figure 2-1).

Figure 2-1
Constant speed rotation of the vector creates a sine wave.

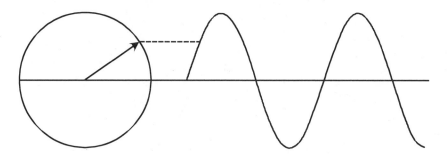

As the wheel rotates, the marker moves up and down, and if we plot its height against the rotation of the wheel (or against time, if the wheel is rotating at a constant speed), we obtain a sine wave. Let's examine the effect of sampling this waveform at various rates.

If we sample at a fairly high rate (16 times per complete revolution in Figure 2-2), we generate a set of samples (Figure 2-3) that clearly is adequate to reconstruct the original accurately (Figure 2-4).

However, it is possible that the movement of the wheel is not regular, and contains higher-frequency components. Conceivably, we could have a waveform like Figure 2-5.

Figure 2-2
Sampling at 16x the frequency of the sine wave.

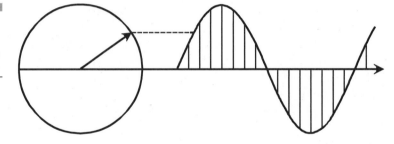

Figure 2-3
The samples resulting from the process of Figure 2-2.

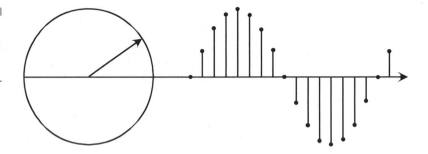

Figure 2-4
Reconstruction of the samples of Figure 2-3.

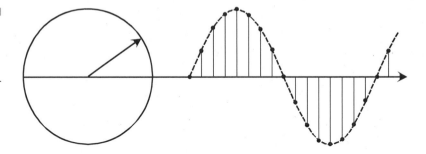

Sampling the waveform in Figure 2-5 generates exactly the same set of samples as we saw in Figure 2-3. The reconstruction will obviously be the same as Figure 2-4; all of the high-frequency information will be lost.

It is easy to see that to capture the higher-frequency information in Figure 2-5, we would need a higher sampling rate; equally obviously, for any given sampling rate we can generate a waveform that contains frequencies that cannot be reproduced with that sampling rate. As we said at the beginning of the chapter, we can choose a precision, and by appropriate sampling and reconstruction we can reproduce the original *to that degree of precision.*

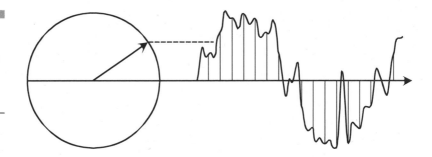

This waveform contains frequencies that will not be captured by this sampling frequency.

If loss of high-frequency information were all that happened when the sampling rate is too low, we would not have a big problem. The sampling process would be acting as a filter, and we would know that the high-frequency information would somehow disappear. Unfortunately, it is not that simple.

Sampling is a multiplicative process; the waveform to be sampled is multiplied by the sampling waveform—usually a train of pulses. In the frequency domain, the effect of this multiplication is to modulate one signal with the other, and *sidebands* are generated. If the sampling frequency is f_s, we can diagram the process very simply:

Figure 2-6 shows the base frequencies (those in the waveform being sampled) and the first set of sidebands. Theoretically, with a perfect point sampling, there is an infinite sequence of sidebands about each integer multiple of the sampling frequency. However, most of the energy is in the first set of sidebands, and in practice these are usually the only ones we need to consider.

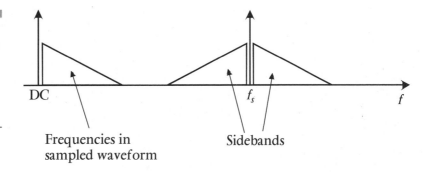

For input frequencies less than half the sampling rate, baseband and sidebands may be separated by filters.

In the case shown, the frequencies in the sampled waveform are quite low, and are well separated from the sidebands. When we come to reconstruct the waveform, we need a filter that passes all the base frequencies, and stops all of the first sideband and everything above. This is not difficult to achieve.

In Figure 2-7, we show the result of sampling a larger range of frequencies. Here we can see that there is no possible filter that can separate the base frequencies from those of the first sideband. In fact, we can see that a base frequency f_1 will cause a frequency component $(f_s - f_1)$ in the first sideband. If the baseband and sideband frequencies are separated as in Figure 2-6, we can exclude sideband frequencies from the reconstruction. If, however, there is an overlap, as in Figure 2-7, and the frequencies fall into the shaded area, there is no way to tell whether a frequency was part of the original signal, or part of the sideband. In fact, if f_1 is greater than $0.5f_s$, the frequency $(f_s - f_1)$ will appear in the reconstruction, and there will be no way to tell that $(f_s - f_1)$ was not part of the original signal. The frequency $(f_s - f_1)$ is an *alias* of f_1 in the reconstruction.

Figure 2-7

When the input contains frequencies more than half the sampling frequency, baseband and sidebands overlap, causing aliasing.

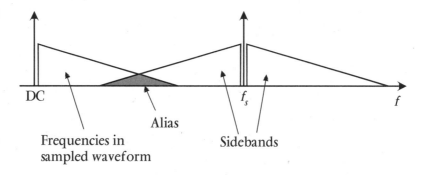

DC

Alias

f_s

f

Frequencies in sampled waveform

Sidebands

Charles Poynton provides an excellent example in his *Technical Introduction to Digital Video* (Poynton, 1996); Figure 2-8 shows sampling (at f_s) of a frequency of $0.65f_s$. It is easy to see that the samples will be reconstructed to yield the alias frequency $(f_s - 0.65f_s)$ or $0.35f_s$.

Suppose we sample the behavior of an analog clock by looking at the position of the hands at regular time intervals. For each hand, we will sample a single parameter, the angle of rotation, at various discrete values of a single variable, time. What we see depends on when we look, but for simplicity and symmetry we'll assume that the hands start at the 12 o'clock position every time we start sampling. If we sample every 5 minutes, we generate a sequence of samples, and see that the rotation of the hour hand increases by 2.5° every sample, whereas the minute hand progresses in the same direction, but at 30° per sample. Having defined *clockwise* as the direction in which we measure

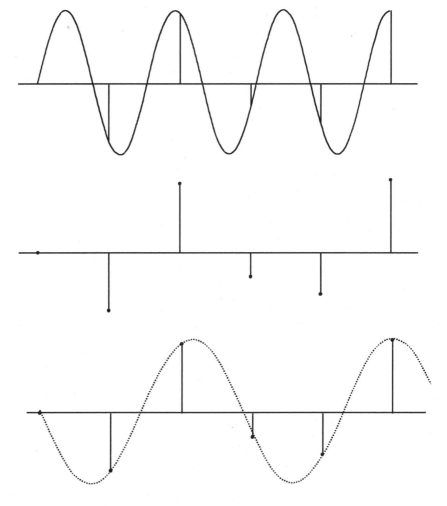

increasing angles, we can conclude that both hands are rotating clockwise, one at 12 times the rate of the other.

If we sample every 15 minutes, or every 25 minutes, we reach the same conclusion. But suppose we sample every 30 minutes? We still have no problem with the hour hand, but the minute hand is at 12 o'clock and 6 o'clock on alternate samples. We may still assume from our knowledge of the system that it is rotating, but which way? If we were to sample every 60 minutes, we would conclude that the minute hand was not moving at all, because it would be at 12 o'clock every sample.

If we sample every 45 minutes, there is no *apparent* ambiguity. The hour hand is progressing clockwise at 22.5° per sample. The minute hand, however, is seen at successive samples to be at 12, 9, 6, 3, and 12 o'clock. It appears to be moving *counterclockwise* at 90° per sample! It seems to be moving at 4 times (not 12 times) the speed of the hour hand, and in the opposite direction.

What have we learned? If we express the movement of the minute hand as a periodic motion with a frequency of one cycle per hour, we are able to analyze its motion correctly for every sample rate above 2 cycles per hour (a sample interval of less than 30 minutes). At exactly 2 cycles per hour there is an ambiguity; below 2 cycles per hour any reconstruction of the minute hand's motion is definitively wrong. We have not only lost information, we have generated false information. This effect is called *aliasing*, and the best definition of aliasing I have heard is "one frequency masquerading as another." In this case a clockwise rotation at 360° per hour appears as an counterclockwise rotation of 90° per hour.

The rule we have observed for the clock is quite general, and known as the *Nyquist theorem*. The theorem can be expressed as:

> To ensure that a signal can be recovered from a series of samples, the sampling frequency must be more than double the highest frequency in the signal.

There is, of course, a very familiar example of the phenomenon of aliasing. A movie-film camera records (samples) a scene 24 times per second. If there is a slowly moving wagon in the scene, the spokes of the wheel will be going by at less than 12 spokes per second, i.e., less than half the sampling frequency. The spokes are correctly sampled according to Nyquist and when we watch the movie we perceive the wheels to be rotating normally. If, however, the wagon speeds up, above about 10 miles per hour, the spokes are going by at more than 12 spokes per second. The Nyquist theorem is violated and, as in the example of the clock, we perceive the wheels to be going backwards. Figure 2-9 illustrates this effect.

Note that in the application of the theorem, there is nothing special about time. The example of the clock used a parameter (the angle of rotation) that changed with time, and so the sampling frequency had to be expressed as time. But if we need to sample a film image with the highest (spatial) frequency information being n lines per millimeter, Nyquist still applies. The theorem tells us that we need a spatial sampling frequency of more than $2n$ samples per millimeter. (In this case, to capture all the information we would need more than $2n$ samples/mm in two dimensions, usually horizontal and vertical.)

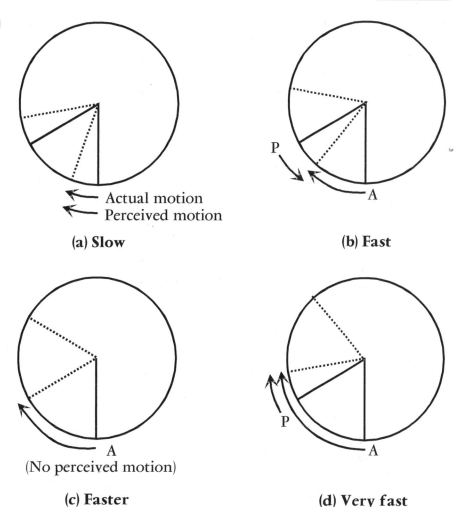

Figure 2-9
Apparent motion of a wagon wheel. For simplicity, only two spokes are shown, widely spaced. (a) For slow motion, the angle moved between temporal samples is less than half the angle between spokes; correct motion is perceived. (b) When the angle per sample is greater than half the angle between spokes, the brain assumes that the shorter distance has been moved, and "sees" reverse motion. (c) When the angle per sample equals the angle between spokes, the wheel appears stationary. (d) At higher speeds the perceived direction is correct, but much slower than actual.

⟵ Actual motion
⟵ Perceived motion

(a) Slow

(b) Fast

(No perceived motion)

(c) Faster

(d) Very fast

For a given sampling frequency, Nyquist tells us that the maximum signal frequency that can be reconstructed is half the sampling frequency, and this is known as the *Nyquist frequency*. However, our analysis of the clock shows us that if the signal frequency exceeds the Nyquist frequency, not only do we lose information, we generate false information or aliases.

Once an alias is generated, there is no way to tell whether it is an alias or a valid part of the original signal. In the example shown in Figure 2-8, if the sampled signal contains a component at $0.35f_s$, there is no way of knowing whether this component existed in the original signal, or if it is an alias of a component of the original signal at $0.65f_s$, or some mixture of both. It follows that once an alias has been generated, there is no way of separating it from the "real" signal, so we must make sure to avoid aliasing. The only way to avoid aliasing is to remove frequencies that will cause aliasing before the alias is created, and it is created by the sampling process. That is why filtering is such an important part of analog-to-digital conversion; to avoid aliasing, frequencies above the Nyquist frequency must be removed *before* sampling. This type of filter is known as an *anti-alias filter*.

It is worth looking briefly at another interpretation of the Nyquist limit that may make it easier to visualize. If we have a waveform and a set of samples, the waveform can be reconstructed from the samples only if the original waveform is the simplest continuous curve that will join all the sample points. Any more complex curve will contain frequencies above the Nyquist frequency; it will contain information that we will "miss" because the samples are too far apart.

Two-Dimensional Sampling

If we have an image, rather than just a waveform, we need to sample in two dimensions, along two axes usually designated x and y. Generally, the image can be represented by the smallest number of samples if the two sampling axes are orthogonal. Usually we choose horizontal and vertical. For most practical purposes the two axes can be considered independent, and the sampling can be performed in either order. We can choose a value of y and then measure the intensity of the image for each value of x. This process is repeated for each value of y. Alternatively, we can choose a value of x and sample the intensity for each value of y, then repeat for the other values of x. The two processes are equivalent and result in the same set of samples.

Even in a static image, the relationship between samples taken in two directions is quite complex. Consider a white bar, which we see as two changes in intensity as we sample in the direction of increasing x (Figure 2-10).

If this bar is the same for all values of y (in other words, it is the same from top to bottom of the image), samples we take in the y direction (for any fixed value of x) will show a constant amplitude for all values of y (Figure 2-11). If, however, the same bar is in a different x position for each value of y (Figure 2-12), then samples taken in the y direction will show changes (Figure 2-13). When a change in amplitude

Figure 2-10
*Samples in the x
direction for a single
value of y. The
intensity increased,
then decreases, as
we move in the
direction of
increasing x.*

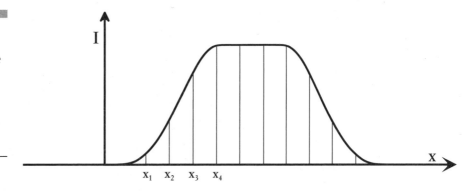

Figure 2-11
*When the bar is in
the same position
for each value of y,
no information is
generated in the
y direction. (For any
given value of x,
intensity is constant
for all values of y.)*

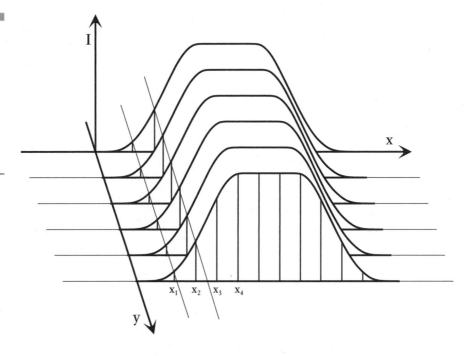

with respect to x has a different x position for each y, then information is seen when we sample in the y direction.

To avoid aliasing, we must obey the Nyquist theorem for any sampling direction. If an image is sampled with the y samples farther apart than the x samples, perfectly legal x information can generate illegal y information. As an example, look at the tilted bar in Figure 2-14. The changes in the x direction are correctly sampled according to Nyquist, but the spacing of the y samples is too great, and aliasing occurs.

Figure 2-12
The changing
position of the bar
in the x direction
generates
information in
the y direction.

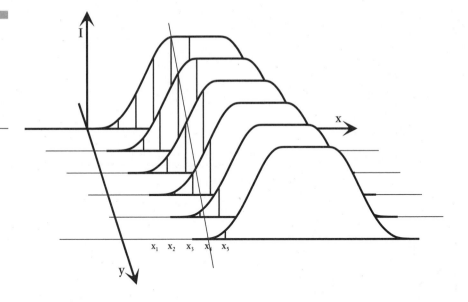

Figure 2-13
Intensity values from
Figure 2-12 for
different y when x
is a constant x₄.

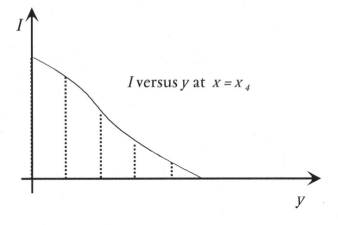

I versus y at $x = x_4$

Generally in image processing the sampling grid is square, or
approximately so. In other words, the samples in the x direction are
spaced the same, or nearly the same, as those in the y direction. This
does not guarantee freedom from Nyquist violations, but in practical
applications problems rarely arise within the static image. Because the
same units of measure are used in both directions, the result is reasonably intuitive. As we will see, life is not quite so simple when we consider sampling in the time dimension.

Figure 2-14

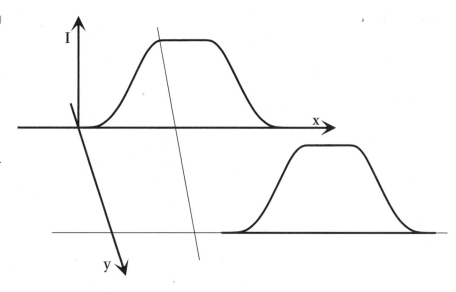

Figure 2-14
This diagram shows the same degree of movement of the bar as Figure 2-12, but there are fewer samples in the y direction. Aliasing is created.

Temporal Sampling

To represent a static image, we must sample it in two dimensions, normally horizontal and vertical. For moving images, we must also sample in time. With motion-picture film, for example, one frame of film is exposed 24 times a second. In video, the photosensitive image area of the tube or charge-coupled device (CCD) is sampled every field (50 or 60 times a second). An understanding of this temporal sampling and its effects is essential to the design of a compression system for moving images.

It is important to remember that when a moving image is sampled, in no sense are we "tracking" moving objects. When a static image is sampled, the amplitude or intensity of each point on the x, y sampling grid is measured. For a moving image system we perform the same process, but repeatedly at different points in time. Just as the x and y sampling points are normally equally spaced, we usually sample at regular intervals in time.

In the case of a static image, we collect a one-dimensional array of intensity samples in one direction for each value in the other direction. That is, for each value of y we collect an array of intensity samples for the various values of x. Or, for each value of x, we collect an array of intensity samples for the various values of y. When either process is complete, we have a two-dimensional array of intensity samples for the required range of x and y.

For a moving image, we repeatedly collect this two-dimensional array of intensity values for various values of time t and arrive at a three-dimensional array of intensity samples for the required range of x, y, and t (Figure 2-15).

So, in sampling an image, what generates information along the time axis? Some things are very obvious—changing illumination, for example, will obviously result in time-varying intensity values. Also, it is fairly easy to see intuitively that a moving object in the scene will generate time-varying information. However, the relationship between the moving x, y information and the resulting time variation of intensity values is somewhat less obvious. We must examine this relationship to understand the characteristics of the temporal information in a sampled moving image, and we must understand these characteristics to be able to compress moving images.

Figure 2-15
Each sample in time (t) is a two-dimensional array of intensity samples (I) in x and y.

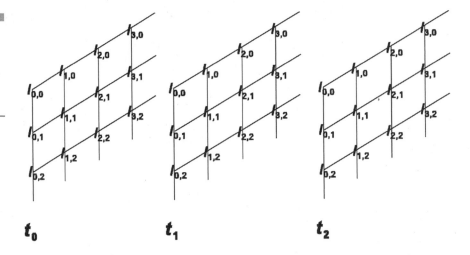

t_0 t_1 t_2

In the previous section we looked at a change in intensity with x, and what happened when the x position of the change varied with y. We saw that this generated information in the y direction. Exactly the same happens in the temporal domain. Figure 2-16 is a replica of Figure 2-14, except that y has been replaced by t. If the edge moves slightly between temporal samples, successive samples will represent points on the edge. The waveform representing the variation in time of the intensity at one point will be validly sampled. If, however, the movement of the edge is enough that the next temporal sample "misses" the edge, we have no way to determine where the edge has gone. We can no longer construct the waveform from the samples; we have violated Nyquist.

Figure 2-16

A replica of Figure
2-14, except that y
is replaced by t.
Temporal aliasing
is created.

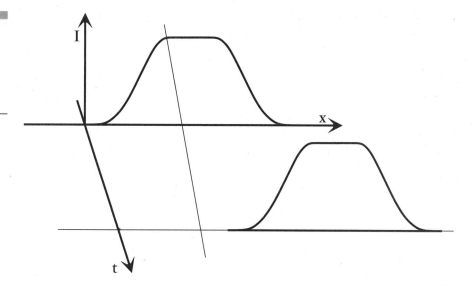

This demonstrates an effect that is true in all dimensions, but particularly important in temporal sampling. A sharp edge (spatially) will generate temporal aliasing at a lower rate of motion than a soft edge. We can perform some crude calculation to see where we might stand with a video system. The sharpest permissible edge is one that moves full excursion between adjacent samples. This corresponds to a frequency equal to the Nyquist frequency, half the sampling frequency. (This is actually the point of ambiguity—we only obey Nyquist if the frequency is slightly less than this, but we will stay with this value for simplicity of calculation.) To avoid temporal aliasing, the same result must apply between temporal samples, so the edge must not move by more than one spatial sample between two temporal samples. In the case of standard-definition digital television, there are 720 samples per line and 60 samples per second. The fastest permissible motion for a sharp edge is that which travels from one side of the screen to the other in 12 seconds! Even an edge with four times this risetime generates temporal aliasing if it traverses the screen in less than 3 seconds.

At first this seems a terrible result. It implies that all current motion imaging systems suffer from gross temporal aliasing. Actually, this is true, and as we will find in later chapters, this has a profound effect on the compression of motion sequences. However, the situation is not as bad as it might appear, for two reasons.

The first is just luck. The human psychovisual system is remarkably tolerant of temporal aliasing. When the eye receives a sequence of images, the brain does an excellent job of applying its knowledge of the universe to the sequence, and "understanding" how the objects in the scene are actually moving. It can certainly be fooled (the wagon wheels are an excellent example), but in general the brain generates an acceptable interpretation of an aliased sequence.

The other reason we get away with low-rate temporal sampling is blurring. In the practical world we cannot sample instantaneously. In a film or television camera each frame requires a certain exposure time. In tube cameras, the exposure time is almost exactly equal to the sampling period, because there is no shutter. Film cameras have a shutter that exposes the film frame for some part of the 1/24-second sample period. (Film exposures are expressed in degrees, where 360° represents the full sampling period; a typical value might be 150°.) CCD television cameras have shutters that can be adjusted over a wide range.

The effect of the exposure time is blurring—any fast-moving detail moves some significant distance during the exposure time, and the effect is to turn sharp edges into soft edges, thus making them less liable to temporal aliasing. The effect is not sufficient to remove all temporal aliasing, but in conjunction with the human psychovisual system it is usually adequate to provide an acceptable rendition of a moving scene, provided the limitations of the system are known. A cinematographer will avoid rapid pans or zooms because these cause objectionable aliasing (seen as judder) at the 24-Hz film sampling rate.

The sampling aperture, or exposure time, acts as a filter, removing some of the frequencies that cause aliasing. (This is just another way of expressing the fact that sharp edges get soft.) The filter is rather crude, having a frequency response that certainly does not remove all aliasing frequencies, but it is much better than nothing. Charles Poynton has an excellent discourse on sampling apertures and the corresponding responses in his *Technical Introduction to Digital Video* (Poynton, 1996), and I refer the reader who wants to know more to this source.

There are two interesting examples that demonstrate the beneficial effect of blurring or, more accurately, the disadvantages of its absence. Artificial imagery such as that generated by computer rendering systems can have sharp edges, and in the early days of this art form, this was the practice. Eventually, practitioners realized that deliberately softening moving edges to provide an equivalent to exposure time blur provided a much more acceptable result. Television, unfortunately, has moved in the opposite direction. The improved sensitivity of CCD cameras allows them to be used with very short exposure times, and directors cannot resist using this capability, despite the obvious artifacts (like gymnasts with many arms) that are produced.

The important thing to remember when considering compression systems is that motion sequences from practical systems carry substantial temporal artifacts. Remember Figure 2-9; when we look at a sequence representing a wagon wheel, there is no way to determine which spoke is which or to determine unambiguously what motion has occurred. We return to this issue in Chapter 8.

Entropy Coding

Introduction

In Chapter 1 we looked at the concept of lossless coding or lossless compression. In the lossless domain, any coding or compression can be entirely and exactly reversed—the bit stream at the output will be exactly the bit stream we started with. Lossless compression can be used alone for any data stream, and we will look at some lossless compression schemes for images. In video, however, lossless compression is more usually used after lossy compression to encode the resulting data as efficiently as possible.

Entropy

Entropy is a measure of disorder, or unpredictability. In information systems, the degree of unpredictability of a message can be used as a measure of the information carried by the message. This may seem somewhat strange to television engineers—we equate disorder with noise, and the part of a signal that is *not* noise is the information. The important concept is that of *predictability*; a perfectly predictable message conveys no information. If we know what a message is going to be, our store of information is not changed by receiving it. If the message is to some degree unpredictable, we know more after receipt than before.

In 1948, Shannon defined the information conveyed by an event *I(E)*, measured in *bits*, in terms of the probability of that event *p(E)*.

$$I(E) = \log_2\left(\frac{1}{p(E)}\right)$$

[handwritten annotation: The great p(E) the less I(E)]

[handwritten annotation: More predictable less information]

We can see that this is at least a reasonable proposition. The higher the probability of an event (i.e., the more predictable that event), the less information is conveyed by that event. A totally predictable event (*p = 1*) carries zero information ($log_2(1) = 0$), as discussed in the previous paragraph. Also, if we look at a sequence of events, the probability of a particular sequence is the *product* of the individual probabilities of the events that make up the sequence. However, the information conveyed by the sequence is the *sum* of the information conveyed by each event, so a log function is expected.

In discussing entropy, theorists use the concept of a *discrete memoryless source* (DMS). A DMS is a generator that sends out symbols, one at a time. The symbols are from a known set or *alphabet*; for example, a DMS might generate symbols that are all digits in the range 0 to 9. Its *alphabet* is the digits 0 to 9. The generator is called *memoryless* because it is stated that the probability of any symbol being generated is independent of what went before. An honest die is a DMS; every time it is thrown it generates a symbol between 1 and 6, and the probability of a number appearing does not depend on the numbers that have come up before.

A DMS need not have the same probability for each member of its alphabet. For example, a *pair* of dice is a DMS; the probability of a value is not influenced by previous throws. However, the probability of throwing a seven (which is created by several different combinations of numbers on the individual dies) is considerably greater than the probability of throwing a two or a twelve (each of which results from only one combination). We might create a DMS that generated the letters *a* through *d*, but such that the letter a is 7 times more likely than *b*, *c*, or *d*. With such a DMS we can list the probability (and, therefore, information content) of each possible symbol:

E	p(E)	I(E)
a	0.7	0.515
b	0.1	3.322
c	0.1	3.322
d	0.1	3.322

The *entropy* of an information source such as a DMS is defined as the average amount of information conveyed by each symbol output by the source. Because the probability of each member of the alphabet determines its frequency, the average information per symbol is the sum of the information in each member of the alphabet, multiplied by its probability.

If the source has an alphabet of n symbols, s_1 to s_n, and the probability of member s_i is $p(s_i)$, then the entropy $H(S)$ of the source can be calculated:

$$H(S) = p(s_1) I(s_1) + p(s_2) I(s_2) + \cdots + p(s_n) I(s_n)$$

normally abbreviated to:

$$H(S) = \sum_{i=1}^{n} p(s_i) I(s_i)$$

which equates to:

$$H(S) = \sum_{i=1}^{n} p(s_i) \log_2 \left(\frac{1}{p(s_i)} \right)$$

This may seem very esoteric, but it is very important when we come to look at coding information efficiently. Look again at the example above of the DMS with alphabet *a* through *d*. Instead of the probabilities given, let us first assume that all members of the alphabet are equally likely. In other words, each has a probability of 0.25. The information content of each member of the alphabet is also the same:

$$I(E) = \log_2 \left(\frac{1}{0.25} \right) = \log_2 (4) = 2$$

so the entropy of the source is the summation of probability multiplied by information for each alphabet member:

$$H(S) = 0.25(2) + 0.25(2) + 0.25(2) + 0.25(2)$$
$$= 2 \text{ bits/symbol}$$

whereas the original probabilities would give us a different entropy:

$$H(S) = 0.7(0.515) + 0.1(3.322) + 0.1(3.322) + 0.1(3.322)$$
$$= 1.357 \text{ bits/symbol}$$

What does this really mean? According to Shannon's *noiseless source encoding theorem*, if we code a source in the most efficient manner possible, and the code is uniquely decodable (no ambiguities), the average number of bits per symbol used by code must be at least equal to the entropy of the source.

Now, we have said that our source has an alphabet of four symbols (*a* through *d*). How would we expect to code this? Given four possibilities, it is not difficult to decide to use 2 bits for each symbol; we have the codes 00, 01, 10, and 11 for the four members of the alphabet.

If we compare this with entropy, in the case of equal probability this code exactly matches entropy (2 bits/symbol), and Shannon's theorem tells us that no more efficient code is possible. However, when the members of the alphabet do not have the same probability, the entropy is lower than 2 bits/symbol, and Shannon suggests that we may be able to find a more efficient code.

How do we average better than 2 bits per symbol? We cannot use fractional bits in the code, so obviously at least one member of the alphabet must be represented by less than 2 bits, and the only possibility is a 1-bit code. If we want the lowest possible average bits per symbol, clearly we must use the shortest code for the most frequent symbol, in this case the letter *a*, which occurs 70 percent of the time. The only possible 1-bit codes are 0 and 1, so let us arbitrarily choose the code 0 to represent the letter *a*. What codes do we use for the other letters? Obviously, the codes must all start with 1, because we would interpret a leading 0 as an *a*. So, a code beginning with 1 represents *b*, *c*, or *d*. The next bit must be a 0 or 1; let's choose 0 to represent *b*, and 1 to represent either *c* or *d*. We then need a third bit to differentiate between *c* and *d*; let's choose 0 for *c* and 1 for *d*. This gives us the following code table:

Symbol	Code
a	0
b	10
c	110
d	111

Is this really more efficient? In a typical group of 10 symbols, the probabilities tell us that the letter *a* will appear seven times, using 7 bits; *b* will appear once, using 2 bits; *c* and *d* will each appear once, and each use 3 bits. The 10 symbols use 15 bits in all, or 1.5 bits/symbol. Not as good as Shannon tells us we might be able to achieve, but 25 percent more efficient than the obvious code of two bits per symbol.

Can we decode the string that would be produced? If we encode the sequence:

acdaaabadaaaaaaacbaa

we get:

01101110001001110000001101000

How do we know which bits belong to which symbol? It's actually quite easy; this code is what is known as a *prefix condition code*. In other words, no code is a prefix to another code. We just have to start at the beginning, and separate off any group of bits that appears in the code table. The first zero is separated, because 0 is in the code table (and, because this is a prefix condition code, 0 is not the start of any other codeword). Neither 1 nor 11 appears in the code table, but 110 does, so we separate it. If this process is continued, the whole string is parsed correctly:

0/110/111/0/0/0/10/0/111/0/0/0/0/0/0/0/110/10/0/0/

We have created an example of a *variable-length code*. We saw a simple example of this in Chapter 1, in Morse code for the English language. We'll look at the most common method of generating a variable-length code in the next section, and at some more sophisticated techniques for obtaining maximum efficiency.

Why was this code not even more efficient? Entropy was 1.357; we only got down to 1.5 bits per symbol. How would we recognize an efficient code? If we assume for a moment that we could use fractions of bits, we can see the answer to both questions. Suppose we were able to encode each member of the alphabet with the number of bits of information it represented. We would use 0.515 bits for *a*, and 3.322 bits each for *b*, *c*, and *d*. The calculation of the average number of bits per symbol would be the same formula as that for the entropy; the average bits per symbol would be exactly the same as the entropy.

This also reveals the inefficiency of the code we created; ideally we should have used just over half a bit to represent the letter *a* (0.515 bits). Encoding one symbol at a time, we can never use less than one bit for a symbol, so we cannot reach the theoretical maximum efficiency.

Before we leave our 4-bit alphabet, we'll examine one more probability distribution that does yield an efficient variable-length code. This also gives us a clue about how to build efficient codes for larger alphabets and complex probability distributions.

E	P(E)	I(E)
a	0.5	1.0
b	0.25	2.0
c	0.125	3.0
d	0.125	3.0

In this very artificial example, each member of the alphabet has an information content that is an integer number of bits. The entropy of the source is 1.75 bits/symbol. We can construct a code table very easily; in fact it is the same we used before:

Symbol	Code
a	0
b	10
c	110
d	111

In this case, however, a typical sequence of 40 symbols contains the letter a 20 times (20 bits); the letter b 10 times (20 bits); and c and d each 5 times ($5 \times 2 \times 3 = 30$ bits), a total of 70 bits for 40 symbols, or 1.75 bits/symbol, exactly the same as the entropy.

Huffman Codes

In 1952 Huffman demonstrated a means for constructing compact (efficient) variable-length codes, and this method remains the most commonly used today. It is the technology behind the lossless coding used in both the JPEG and MPEG compression standards that are discussed in later chapters.

The construction of a Huffman code involves two processes: source reduction followed by *codeword construction*. Each process has several steps, depending on the size of the source alphabet.

In the source reduction process, the two least-probable symbols are combined, leaving an alphabet with one less symbol. The combined symbol has a probability equal to the sum of the probabilities of the two original symbols. When we allocate codewords, we decide on a codeword for the combined symbol, then add 0 or 1 to it to generate codewords for the two original symbols. The source reduction process continues; at each step we combine the two least-probable symbols and create a combined symbol with summed probability. Eventually, there will be only two symbols, and the first step is complete.

We'll demonstrate this process with a new source whose alphabet is s_1, s_2, s_3, s_4, s_5:

E	P(E)	I(E)
s_1	0.30	1.737
s_2	0.25	2.000
s_3	0.20	2.232
s_4	0.15	2.737
s_5	0.10	3.322

First, we will go through the source reduction process. The symbols are arranged in order of priority, so the first step is to combine the two lowest symbols. At each step we will designate a combined symbol by combining the suffixes of its components. For example, the combination of s_4 and s_5 is s_{45}—using parentheses as necessary for clarity. The term s_{45} means s_4 *or* s_5, and has a probability equal to the sum of the probabilities of s_4 and s_5. After each combination, the new set of symbols is reordered according to the new probabilities, so that again we can combine the two lowest symbols.

Original source alphabet		Reduced source stage 1		Reduced source stage 2		Reduced source stage 3	
s_i	$p(s_i)$	s'_i	$p(s'_i)$	s''_i	$p(s''_i)$	s'''_i	$p(s'''_i)$
s_1	0.30	s_1	0.30	$s_{3(45)}$	0.45	s_{12}	0.55
s_2	0.25	s_2	0.25	s_1	0.30	$s_{3(45)}$	0.45
s_3	0.20	s_{45}	0.25	s_2	0.25		
s_4	0.15	s_3	0.20				
s_5	0.10						

As stated, source reduction is performed until only two symbols remain, in this case s_{12} and $s_{3(45)}$. At this stage, code allocation is trivial; with only two symbols, we allocate the codes 0 and 1. Conventionally, at each step we use 0 to represent the higher priority and 1 the lower priority. If both are 0.5, the choice is arbitrary. Now we split each composite symbol by appending 0 and 1 to the code of the composite symbol, and we continue this process until all the composite symbols have been reduced to original symbols of the alphabet. We now have a code for each member of the alphabet.

Code allocation stage 3		Code allocation stage 2		Code allocation stage 1	
s_i	*Code*	s_i	*Code*	s_i	*Code*
s_1	00	s_1	00	s_{12}	0
s_2	01	s_2	01		
s_3	11				
s_4	100	s_3	11		
s_5	101	s_{45}	10	$s_{3(45)}$	1

Note that this is not the only possible code; at each step where we assigned 0 and 1 to two symbols, we could have reversed the allocation without any change to the efficiency of the code. For each symbol set, therefore, there exists a number of possible Huffman codes having equal efficiency. It can be shown that it is not possible to generate a code that is both uniquely decodable and more efficient than Huffman (Abramson, 1963).

Dangers of Variable-length Coding

A properly constructed variable-length code is efficient because it is derived from the probability distribution of the symbols. This is fine, provided we do indeed know the probabilities and they do not change. However, the very reason that a code is an efficient code means that if the probability distribution does change, the code efficiency will be lost. When we looked at the Morse code in Chapter 1, I emphasized that the code was modeled on the probability distribution of the letters in English-language text. We saw that the code would not be efficient if used for Polish—in fact it would be less efficient than a code that assumed equal probability for all symbols.

Returning to the four-symbol alphabet, we know that we could code this source with a fixed length code of 2 bits for each symbol, and this yields an average of 2.0 bits/symbol, irrespective of the probability distribution of the alphabet. When we examined the source with probabilities 0.7, 0.1, 0.1, 0.1 we were able to construct a code that used an average of 1.5 bits per symbol—a 25 percent improvement over the fixed-length code. Now let's see what happens if the probability distribution changes.

We will assume that *a* and *d* swap probabilities, so now we have 0.1, 0.1, 0.1, 0.7. If we use the same variable-length code derived from the original distribution, we can again calculate the bits used by a typical group of 10 symbols. The letter *a* will appear once using 1 bit; *b* will appear once using 2 bits; *c* will appear once using 3 bits; and *d* will appear seven times using 21 bits. The 10 symbols use 27 bits in all, or 2.7 bits/symbol, 35 percent worse than the fixed-length code.

The lesson is simple; if we are sure of the probability distribution of a source, a variable-length code can offer substantial advantages. If for any reason the assumed probabilities are wrong, or the source distribution changes, the variable-length code can do more harm than good.

Modified Huffman Codes

In the discussion of Huffman codes we have considered examples with very small alphabets. These are useful for demonstrating the principles involved, but obviously such small alphabets are not typical of the real world. In image processing the sources usually produce a much greater range of values. Typical ranges include 0 to 255, -254 to $+255$, and -510 to $+511$. As we will see in the next chapter, lossy coding schemes often use quantization to reduce the number of possible values, but at the moment we are considering lossless techniques, and we must be prepared to encode any possible value in the range.

If the probability distribution is known, it is possible to construct a Huffman code for any number of values. However, having seen the method, we know that such a construction would be very laborious. Also, we have to transmit the code table, and that in itself represents a very large amount of data. After all that effort, if the assumed probability distribution is wrong for any part of the alphabet, that part of the code will reduce rather than improve efficiency.

In fact, there is very little benefit in trying to construct such a code. If the probability distribution is fairly even over the full range of possible values, there will be little to gain from the use of a variable-length code. Fortunately, many of the sources we must deal with have very uneven probability distributions. Sometimes this is a property of the data; in other cases we manipulate the data to achieve this sort of distribution.

To demonstrate the principle, I again assume a source with rather exaggerated characteristics. The source has a range of -510 to $+511$ requiring 10 bits/symbol with a fixed-length code. The source, however,

is highly biased toward low absolute values. The probability table looks
like this:

E	P(E)	I(E)
0	0.20	2.232
+1	0.10	3.322
−1	0.10	3.322
+2	0.05	4.322
−2	0.05	4.322
All other values combined	0.5	1.000

The entropy value of 1.0 for "all other values" means that we obtain 1
bit of information from the fact that the value is not one of the five
values listed; the information conveyed by the actual values is not
shown. This table shows that we can derive a variable-length code for
the six entries shown, but how do we code the remaining values? It's
really quite simple—we take the code for "all other values," and append
the actual 10-bit value that must be transmitted. Let's see how this works
in practice. Using the techniques already described, we can derive a
Huffman code for the six entries in the table. With the conventions we
have used, we represent "all other values" by 0, and each of the five most
probable values by a code beginning with 1. As discussed above, we then
append the actual value (10 bits) to the 0. Here is a possible code table:

E	Code	Bits
0	10	2
+1	110	3
−1	1110	4
+2	11110	5
−2	11111	5
All other values combined	0 plus value (10 bits)	11

Now if we consider the typical group of 20 symbols, 0 will appear four times for a total of 8 bits, +1 and −1 will each appear twice for a total of 14 bits, +2 and −2 will each appear once for a total of 10 bits, and there will be 10 "other values" for a total of 110 bits. That gives a grand total of 142 bits, or an average of 7.1 bits/symbol. This very simple code achieves a coding gain of nearly 30 percent; it uses only a very small code table, and we do not have to worry about the statistics of the vast majority of values.

We should make certain that we can decode such a bitstream without ambiguity. We can encode a short sequence of values as shown (with punctuation added for clarity).

254 7 2 −19 −1 0 14 1 0

0-0011111110 0-0000000111 11110 0-1000010011 1110 10

0-0000001110 / 10

Now, if we concatenate these codes into a bit stream (without the punctuation) we get:

0-0011111110 / 0-0000000111 / 1111001000010011111010100000000111010

Remembering the rule, we have to examine the bit stream, looking for words that appear in the code table. Starting at the beginning, the only code table entry starting with zero is the "all other" case, so we know we must extract the zero and the following ten bits. We see immediately that the bit after the ten is also a zero, so again we extract that and the following ten bits:

0-0011111110 / 0-0000000111 / 1111001000010011111010100000000111010

Life now becomes a little more interesting because the next bit is a 1. So, we keep accepting bits until we have collected a string that appears in the code table, in this case 11110. Having extracted that, again we have a 0, so we take that and the following ten bits:

0-0011111110 / 0-0000000111 / 11110 / 0-1000010011 / 1110100000000111010

Following the same rules we extract a 1110, a 10, another 0-plus-ten string, and a final 10. The bit stream is now fully parsed,

0-0011111110 / 0-0000000111 / 11110 / 0-1000010011 / 1110 / 10 / 0-0000001110 / 10

and at no point was there an ambiguity. This does not constitute a proof, but the code table does represent a prefix condition code, and it can be shown that all such streams are uniquely decodable (Abramson, 1963). The additional rule of extracting a fixed number of bits after every "all other" code does not violate this principle.

As noted, this source was given exaggerated statistics for the sake of a simple example; most real sources will have a greater number of frequent values. However, we will see that it is often possible to isolate a small number of frequent values and achieve both simplicity and good coding gain by lumping together all other values.

When we come to consider the lossy technique of quantization, we will see that the data from a good quantizer is well suited to subsequent lossless compression by a modified Huffman encoder.

Arithmetic Encoding

Huffman codes and derivatives can provide efficient coding of many sources and, since most of the work was published in 1952, can be implemented without any issues of intellectual property rights. There are, however, some limitations. Huffman techniques generally encode individual symbols. Ideally, the Huffman codeword length for each symbol is the same as that symbol's entropy, but we are confined to codewords with integer lengths so we cannot generally achieve this. Also, no code can be shorter than 1 bit, so we cannot deal efficiently with a very skewed alphabet where one symbol appears perhaps 90 percent of the time. There are techniques that improve the performance of Huffman codes; symbols can be combined into larger blocks, although at the expense of complexity. We will see some of these techniques in the discussion of JPEG in Chapter 7.

We have also seen that Huffman coding relies very heavily on accurate knowledge of signal statistics. There are methods of handling changing statistics (changing coding tables, etc.) but these are clumsy.

There is another class of entropy encoder, known as the *arithmetic encoders*. Based on work originally performed at IBM, these encoders are continuous-flow devices. A sequence of symbols goes into the encoder, and a bit stream comes out. Coding is performed by a large number of binary decisions, and is based upon continually updated probability estimates. By its very nature the encoding adapts to the statistics of the input. Explanation of the arithmetic coding process is beyond the scope of this book, but I mention it because it is an option in the JPEG standard. The process is very efficient; IBM's Q-Coder operates within about 6 percent of optimum. Arithmetic coding is the subject of current patents, but this can be overcome by buying chips from the patent holders or licensees. The QM-Coder used for compression in facsimile applications (the JBIG standard) is available in inexpensive chip form from several vendors.

Predictive Coding

Markov Sources

In the previous chapter we considered the DMS—the discrete memoryless source. As a reminder, this is a source where the probability of a given symbol value is unaffected by the values of previous symbols. Some sources are indeed memoryless. A symmetrical coin, fairly tossed, has a probability of exactly 0.5 for heads and 0.5 for tails (assuming it cannot stay on edge)—irrespective of what has occurred previously.

Many sources, however, are quite different. The probability distribution of values for one symbol can be quite dependent on one or more previous values. Again, we can use English-language text to provide some simple examples. If one letter is *q*, there is a high probability that the next letter will be *u*, a much lower probability of a space (some words end in *q*), but almost zero probability of any other letter. It is also easy to see that the probability in this case is not just dependent on the one previous letter. A *u* may be followed by almost any letter, but the pair of symbols *qu* is most unlikely to be followed by a consonant.

A source where the probability of a particular symbol value depends on previous values is known as a *Markov source*. If the dependence is on the previous value only, the source is known as *first-order Markov*. Dependence on two previous values gives us *second-order Markov* and so on. Photographic images are Markov sources, and we will use this method heavily in developing compression techniques for these images.

It is interesting here to compare photographic imagery with computer graphics. In a photographic image it is highly probable that the intensity of a pixel will be close to that of its neighbors. In simple computer graphics the situation is rather different. There is a fairly high probability that a pixel will have a value identical to that of its neighbor, but if the value is not identical it is difficult to predict what the value will be—it could be any other permissible value. With more sophisticated computer-generated imagery (CGI), great care is taken to render surfaces so that they have real-world or photographic characteristics. Such a process yields a Markov source.

The study of such sources leads to techniques of predictive coding. Before we look at these techniques we'll take a brief diversion that will help to demonstrate why prediction is such a powerful tool in compression.

It is interesting to look at the characteristics of language in terms of prediction. Such an analysis helps demonstrate the difference between data and information. In an environment with a known structure, data are highly correlated—meaning that much of the data represents the structure. If we can separate the information from the structure (decorrelate) we can transmit just the information, and reapply the known structure at the receiving point.

This analysis is the work of Shannon (1948) and Abramson (1963), and some of the examples are quoted from Abramson's book.

As a first step, we assume that we know nothing of the structure of the English language. The best "structureless" starting point is a discrete memoryless source with equal probability for any of the 27 basic symbols (letters plus space). A sequence from such a source might look like this:

zewrtzynsadxesyjrqy wgecijj obvkrbqpozb ymbuawvlbtqcnikfmp
kmvuugbsaxhlhsie m

We have derived a series of "words" where each word is a string of letters ending in a space, but the result does not look much like English or any other language. The next step is to continue to treat the source as a DMS, but to recognize that not all symbols have the same probability. The probability of the basic symbols of the English language was published by Reza in 1961 and is shown in the table on the next page:

Use of this table will, no doubt, make a significant difference. Note that the probability of an *e* is more than two hundred times higher than that of a *z*, and a space is almost twice as probable again. We will still assume a DMS, but use this probability table to construct a typical sequence:

Ai ngae itf nnr asaev oie baintha hyr oo poer setrygaietrwco
ehduaru eu c f t nsrem diy eese f o sris r unashor

It would be stretching things to say that this sequence looks like English, but it certainly looks less unlike English than the first example.

But this is still assuming a DMS, and we have already seen that English is at least a second-order Markov source. Constructing the probabilities for all the possible sequences is possible, but very arduous. Shannon pointed out that we can use ordinary English text as a source of the probability mapping, because it contains exactly the information we need (in fact, it is what we would analyze to determine the probabilities). To generate a first-order Markov sequence by Shannon's

Symbol	Probability	Symbol	Probability
Space	0.1859	*n*	0.0574
a	0.0642	*o*	0.0632
b	0.0127	*p*	0.0152
c	0.0218	*q*	0.0008
d	0.0317	*r*	0.0484
e	0.1031	*s*	0.0514
f	0.0208	*t*	0.0796
g	0.0152	*u*	0.0228
h	0.0467	*v*	0.0083
i	0.0575	*w*	0.0175
j	0.0008	*x*	0.0013
k	0.0049	*y*	0.0164
l	0.0321	*z*	0.0005
m	0.0198		

method, we take a passage of typical English text, and select a starting letter (it is probably valid to select the first letter of a word, as some letters start words more commonly than others). Let's say we start with a *t.* We then skip some text, say three lines, find the first t on that line (or subsequent lines if there is not one on the first line) and take the letter (or space) immediately following it. Here we find an *s,* then skip three lines, look for the first *s,* and pick the letter immediately following it, another *t.* And so on!

It became evident when I started constructing examples that the results depend (as might be expected) on the sort of text chosen as "normal." A text full of long words generates sequences with long words; text with many similar words produces some strings that look similar to those words. After a number of tries I decided to use the opening chapter of this book as my source of "normal" text—at least I have no one else to blame! Here's a first-order sequence.

Tst dule wh tio h ms me cotativial tellom co d
odifus doreng s ffro tanussiose oubjeves

This looks a little more like English. Let's try a second-order sequence. We select a first letter, and take it *and the following letter.* We skip three lines, and take the letter following the first occurrence of the letter *pair.*

> In earyied thents the comm dat a froces ted
> degressigle sysiblentrobable be vides

Third-order becomes tedious, but may be accomplished with the search facility of a word processor. It yields some rather interesting results. The first example is, again, from the first chapter of this book.

> Techniques nothe many charge codemy some ther world
> of repressed of cover serviced to transfer intern...

It is fairly clear that some words get built because they appear many times in the text on a given subject. For a (very) different writing style I tried the Telecommunications Act of 1996.

> Refers the exempt telecompanies but inted other
> to disputes subsidize to commissions facilization...

Again, the repetitive nature of the text comes through, though not as obviously as it may appear. The word *commissions* in this sequence actually picked up letters from various repetitions of *comply, control, commission, provisions, inspections,* and *telecommunications*!

Finally, I tried this technique on the text of *Alice in Wonderland* with this result:

> Picturnal reply this good myself with it help the
> do wered und said ther should hanglifter...

Not, perhaps, something that bears comparison with Mr. Dodgson's own nonsense rhymes, but there are some interesting similarities!

Shannon (1948) took a different approach, and generated random English words. Again, this is a DMS, but the probability of the words themselves was calculated from an analysis of English text. He obtained the sequence:

> Representing and speedily is an good apt or come can
> different natural here he the a in came the to of to expert
> gray come to furnished the line message had be these...

This is little more than a scattering of random words. However, Shannon also examined the probability of one word following another, and generated this sequence, treating the words as output from a first-order Markov source:

The head and in frontal attack on an English writer that the character
of this point is therefore another method for the letters that the time
of who ever told the problem for an unexpected...

As Abramson (1963) points out, "... the sequence is a reasonable
approximation to the sort of output one might expect from a very
excited and quite incoherent speaker." This really is a remarkable
result. In one case we have taken letters, and by feeding random noise
into a statistical model of the language, we have output that looks
very like words. In the other case Shannon took words, and again by
feeding random noise into a statistical model, he produced something
remarkably like a sentence.

The sequences do not contain any information—they really are just
noise, but noise correlated so that it resembles English. Here we have a
great deal of data, structured to resemble English, but carrying no
information.

This helps me imagine a similar process, but in reverse. If we take a
passage of real English text, we have a certain amount of data, containing some information. If we can remove from the data everything
that merely represents structure, in other words de-correlate the data,
we may be able to transmit just the information, with fewer bits.
From our knowledge of the structure we should then be able to recreate the original data from the information-only transmission.

Now we will return to thinking about images, and look for techniques that let us benefit image structure.

Differential Pulse Code Modulation

Differential pulse code modulation (DPCM) is a form of predictive coding
suitable for use with Markov sources. DPCM is not in itself a compression system; if it is used alone, the quantity of data output will be slightly greater than that input. Instead, DPCM is a technique that decorre-
lates data. It is a means of separating information from structure.

We saw earlier in this chapter that English-language text can be
treated as a Markov source, and we would expect that substantial
advantage could be gained by coding it in this fashion. We also saw
that, treated as a DMS, English text had a very skewed probability distribution. So, even ignoring the Markov characteristics, we expect significant coding gain simply by treating English as a low-entropy
DMS—just as Morse did.

Photographic images are different. There may be a skew of probabilities for pixel intensities if the image is overall dark or light, but there is not much advantage to be gained by coding a typical image as a DMS. Rather, we must use the fact that the intensity value of a pixel is likely to be similar to that of the surrounding pixels—much more likely than being wildly different from that of the surrounding pixels.

This is the ideal situation for predictive coding. The principle is very simple. Instead of coding an intensity value, we predict the value from our knowledge of the intensity values of one or more nearby pixels. We rely on the decoder at the receiving end to perform the same prediction process, and to arrive at the same answer. All we need do is to compare our prediction with the actual intensity value, and transmit the error. The decoder will make the same prediction and correct according to the transmitted error value, arriving at the correct intensity value for the pixel.

As I mentioned previously, this in itself does not provide compression. If the pixel intensity values are in the range 0 to 255, our prediction will be in the same range, but the range of possible error values will lie between -255 and $+255$. (We might predict 255 when the correct value is 0, and we might predict 0 when the correct value is 255.) Transmitting a 9-bit error value is certainly not in itself preferable to transmitting an 8-bit intensity value.

The benefit lies in the fact that our prediction is likely to be quite accurate, because of the nature of the image we are encoding. In other words, there is a high probability that the error will be small rather than large. The error value is a 9-bit value, but we now know something about its statistics, so we can take advantage of entropy encoding.

Something else has happened. Although we now expect most of the error values to be small, there is no particular reason to suppose that the error value for one pixel is likely to be the same as the error value for the previous pixel. What has the technique of prediction and error measurement achieved?

At the input to the process we have a set of correlated intensity values. The entropy of the intensity values is high—all possible values have similar probabilities—but there is a high level of dependence between adjacent values. After the process we have a set of error values with little dependence between adjacent values, but individually the values have a high probability of being small. We have transformed a set of high-entropy but correlated values into a set of lower-entropy but less correlated values. If the process were perfect we would have turned a high-entropy Markov source into a low-entropy DMS.

In Chapter 3 we discussed the use of variable-length encoders to reduce the data needed to transmit a low-entropy DMS. What we see here is an example of a technique that is common in compression systems. One tool is used to change the characteristics of the data to make it suitable for compression by another tool.

In the process of prediction we have gained data (the range of possible error values is twice as large as the range of pixel values), but we have reduced the entropy, thus making the data suitable for variable-length encoding.

Low entropy means that the amount of information is reduced. We can see this; the information is now just the error in our prediction, and if our prediction is any good, this should represent less information than the original sequence of intensity values. Because a photographic image has structure, our predictions are to some degree successful. By the process of prediction followed by error measurement we have removed at least some of the structure from the data, leaving mostly information. The information is less than the original data, so we can transmit it with fewer bits.

It is fairly easy to see that the error data will likely fit another model we discussed in the last chapter. So long as our prediction is good or fairly good, the error value will be low or quite low, and in this area there will probably be consistent probabilities for each error value. When our prediction is bad, for example, if we try to predict over an object edge, the error value will likely be high. Because our prediction mechanism has failed, there is no reason to suppose that any large error values are more probable than any others. This is exactly the scenario where the modified Huffman code is most efficient.

Predicting Image Values

We have not yet discussed how to make the prediction. In a one-dimensional sequence of data it is obvious and convenient to take one or more of the values immediately preceding the value we wish to predict. An image, however, is two-dimensional. For any given pixel not on an edge there are adjacent pixels in all directions.

A predictor can use any number of adjacent and nearby pixels, but the complexity of both encoder and decoder increases rapidly when many pixels are used. In practice it is found that simple predictors work well.

Habibi (1971) describes a theoretical analysis and measurements on real images testing predictors that used from 1 to 22 surrounding pixels for prediction. The theoretical and practical analyses both showed a substantial reduction in mean squared error, going from 1 pixel to 2 and from 2 pixels to 3, but little significant change beyond that.

Predictors that use only 1 pixel are called *first-order*, and are necessarily one-dimensional (1-D). Predictors using 2 or 3 pixels are known as *second-* and *third-order*; both are usually two-dimensional (2-D) in that they use data displaced both horizontally and vertically from the pixel to be predicted.

To minimize storage requirements, the range of pixels used for prediction is usually confined to the set shown in Figure 4-1. This assumes conventional scanning, left to right, and top to bottom. The simplest predictor is simply *A*—in other words, we predict that each pixel value will be the same as its left-hand neighbor. We could use *B*, *C*, or *D*, but with the scanning process we have assumed, it is necessary to store the previous line of pixels to have access to these values. If we do store these pixels, we can easily take advantage of a better predictor by using two or three of the values.

Figure 4-1
The pixels (A,B,C,D) most commonly used to predict pixel X.

B	C	D
A	X	

The most obvious second-order, 2-D predictor is

$$\hat{X} = \frac{1}{2}A + \frac{1}{2}C$$

(The "cap" over X means the predicted value of X.)
One useful third-order, 2-D predictor is

$$\hat{X} = A - B + C$$

This is known as the *plane predictor* because, if the values are regarded as vertical coordinates (perpendicular to the image), the prediction is the value that lies on the plane passing through *A*, *B*, and *C*.

Another third-order, 2-D predictor that can give good results is

$$\hat{X} = 0.75A - 0.5B + 0.75C$$

Figure 4-2 shows two reference images and a histogram of each image indicating the distribution of intensity values. Figures 4-3 to 4-6 show the results of the above predictors operating on each of the reference images. The picture shows the error values (shifted to all positive for printing); the histogram shows the distribution of error values and the entropy of the resulting discrete memoryless source.

Note that these histograms of the original images have ten times the vertical scale of the error histograms that follow (top of scale is 2,000 compared with 20,000 on the following pages).

"Boats" has nearly 15,000 zero errors (22 percent), and over 70 percent of the errors are within ±5. Unfortunately, there are still 248 different error values, even though many appear only once or twice, so that the entropy is not reduced as much as we might expect. Nevertheless, with appropriate coding we could transmit either of these images losslessly with 6 bits per pixel or less, a gain of 25 percent. There are many applications where this is a meaningful improvement.

This predictor requires a full line of storage, so its cost of implementation is higher, but the improvement in performance will often justify this cost. The calculation requirements are very low.

The results are quite impressive; both images produce lower entropy errors than any other we have looked at. Note also that the better the predictor, the more similar the entropy results from the two different images. Of course this test is not conclusive, but it does suggest that the predictor will perform well over a wide variety of images.

This predictor yields 77 percent of the errors in the range ±5, and the number of discrete error values is reduced to 190.

The histograms show clearly that, qualitatively at least, one objective has been achieved. Unlike the broad spread of intensity values in the original image, all of the predictors produce a range of error values very heavily concentrated around zero.

The images constructed from the error values illustrate a number of things. The fact that they are predominantly mid-gray (corresponding in these images to an error near zero) shows that the average error is low. That most of the image detail has disappeared shows that decorrelation was successful—in fact, the degree to which the original picture is discernable gives an indication of the remaining correlation in the set of error values, or the degree of failure of the process. (This is true only to a point; if an image were perfectly predictable it would contain no information.)

It is clear from examining these images that different predictors work with varying degrees of success on different images, and on different areas of one image. You can see by examining the error pictures that some predictors do particularly well, or particularly badly, on specific types of image content. For example, the predictor

$$\hat{X} = A$$

fails badly on vertical edges (Figure 4-3). It would be nice if we could pick our predictor according to the local characteristics of the image, to get the best achievable result at all times.

Adaptive Prediction

In a system using adaptive prediction, several predictors are predefined, and the system switches between these predictors according to the characteristics of the data being compressed. There are two approaches to such a system. We can try out a number of predictors and determine which is the best, then tell the decoder which predictor has been used. This method is very flexible, but it requires some overhead to pass to the decoder the predictor choice message. With appropriate design, the overhead can be very small. For example, if we were to consider blocks of 10 pixels and choose among four predictors, the overhead is only 2 bits per block, or 0.2 bits per pixel. This type of adaptation may also be used to among between different quantizers (see Chapter 6).

Alternatively, we may be able to determine the choice of predictor from the data values in such a way that the decoder can make the same choice from the reconstructed values. If we can achieve this, adaptive prediction is free of overhead.

One adaptive predictor chooses one of three algorithms depending on the values of A, B, and C. Two tests are performed for edges. The predictor assumes that, if B is greater than the larger of A and C, there is a falling edge; if B is less than the lesser of A and C, there is a rising edge. In the case of a falling edge, if A is larger than C we assume a vertical edge, and C is used as the predictor. If A is less than C we assume a horizontal edge and use A as the predictor. Similarly, for the rising edge, C is used for vertical edges and A for horizontal. The expression for the predictor is

$$\hat{X} = \begin{cases} \min(A,C) & \text{if } B \geq \max(A,C) \\ \max(A,C) & \text{if } B \leq \min(A,C) \\ A+C-B & \text{otherwise} \end{cases}$$

A little work with pencil and paper may be necessary to visualize the operation. This predictor is proposed for lossless compression in the new JPEG-2 currently in process of standardization. The results are shown in Figure 4-7.

For "Boats" this predictor yields nearly 17,000 zero errors; over 25 percent of the pixels, and now 80 percent of the errors are within ±5. However, the reduction in entropy comes mainly from the lower number of discrete error values, now down to 126.

On the one hand the improvement over the simple predictor is encouraging; certainly the more sophisticated predictors do a better job. With suitable encoding it looks as though we could transmit most 8-bit monochrome images losslessly with about 4.5 bits/pixel—a compression ratio of almost 2:1.

On the other hand, although there is significant improvement from the simplest to most complex predictors, it is clear that we are in an area of diminishing returns. It is likely that the adaptive predictor shown here is close to the practical limit.

Predictive coding is important in a number of ways. We will see in later chapters that this form of lossless encoding is used in sophisticated compression schemes to compress critical data, such as DC coefficients and motion vectors. We will also see in Chapter 6 that the technique can be made much more powerful if some degree of loss is acceptable.

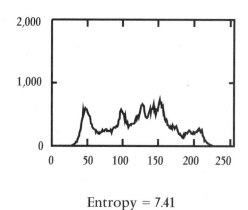

Entropy = 7.41

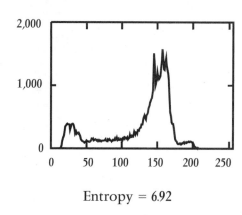

Entropy = 6.92

Figure 4-2   These histograms of the two reference images show the occurrence of the different intensity values in the original images. Both show a reasonably even distribution, with most values appearing at some point in the image. Perfectly even distribution would give 256 pixels of each value. "Lena" peaks at about three times this number, "Boats" at about six times (note the lower entropy)."Lena" reproduced by special permission of Playboy magazine. Copyright © 1972, 2000 by Playboy,

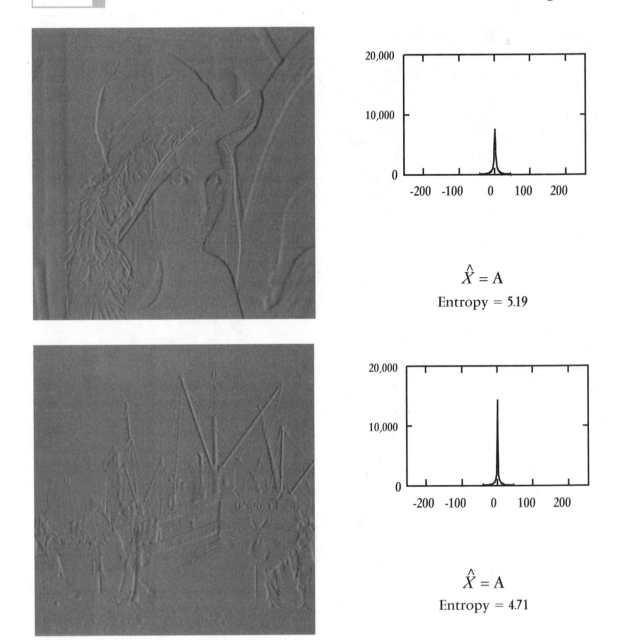

$$\hat{X} = A$$

Entropy = 5.19

$$\hat{X} = A$$

Entropy = 4.71

Figure 4-3 The simplest predictor is "A" (see Figure 4-1), the pixel immediately preceding the pixel being coded. This requires only one pixel of storage at encoder and decoder, and the minimum amount of calculation. Even this simple predictor has a dramatic effect on the distribution of values (note again that the vertical scale on these histograms is one tenth that of Figure 4-2).

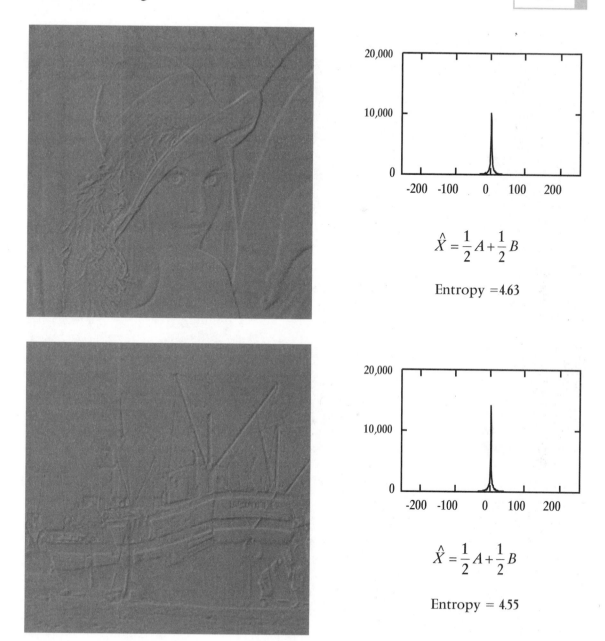

$$\hat{X} = \frac{1}{2}A + \frac{1}{2}B$$

Entropy = 4.63

$$\hat{X} = \frac{1}{2}A + \frac{1}{2}B$$

Entropy = 4.55

Figure 4-4 This example shows the significant improvement that can be obtained with even a simple two-dimensional predictor. This predictor merely takes the mean of the pixel above and the pixel to the left, but the entropy is significantly lower than with the previous one-dimensional predictor, particularly on "Lena."

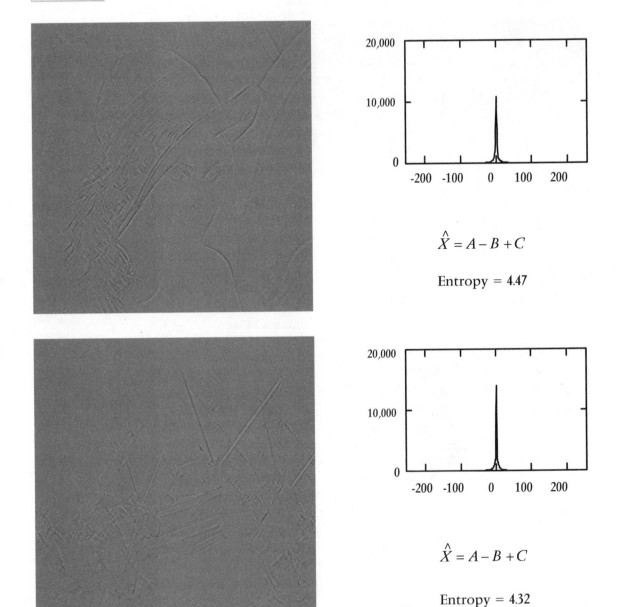

$$\hat{X} = A - B + C$$

Entropy = 4.47

$$\hat{X} = A - B + C$$

Entropy = 4.32

Figure 4-5   This third order, two-dimensional predictor is known as the plane predictor. If the locations of the pixels are taken as x and y coordinates, and their intensity values as z coordinates, X lies on the plane defined by A, B, and C. It uses only one more pixel of storage and a similar amount of computation, but provides somewhat better results on most images.

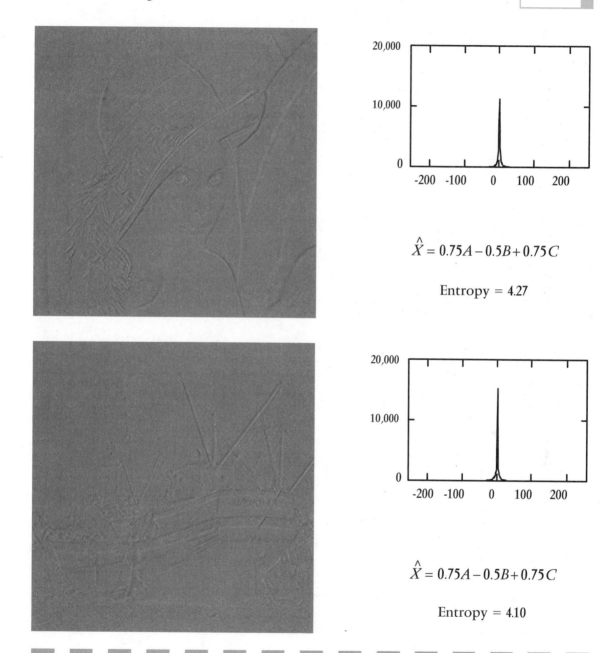

$$\hat{X} = 0.75A - 0.5B + 0.75C$$

Entropy = 4.27

$$\hat{X} = 0.75A - 0.5B + 0.75C$$

Entropy = 4.10

Figure 4-6 This third order predictor uses the same pixels as the plane predictor but manages to combine some of the performance advantages of this with those of the simple A,C predictor shown in Figure 4-3.

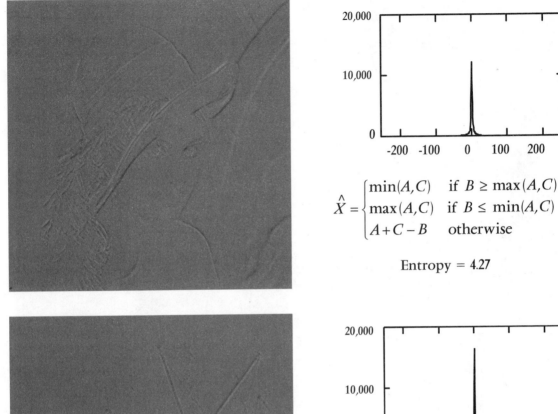

$$\hat{X} = \begin{cases} \min(A,C) & \text{if } B \geq \max(A,C) \\ \max(A,C) & \text{if } B \leq \min(A,C) \\ A+C-B & \text{otherwise} \end{cases}$$

Entropy = 4.27

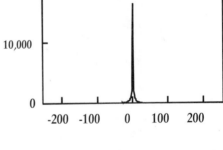

$$\hat{X} = \begin{cases} \min(A,C) & \text{if } B \geq \max(A,C) \\ \max(A,C) & \text{if } B \leq \min(A,C) \\ A+C-B & \text{otherwise} \end{cases}$$

Entropy = 4.10

Figure 4-7   This example uses the adaptive predictor proposed by JPEG-2. Its performance is excellent on both images—indeed it is quite difficult to pick out the details of the original image from the error pictures. This shows that decorrelation has been very successful.

Transforms

Introduction

A transform is a mathematical rule that may be used to convert one set of values to a different set. A transform creates a new way of representing the same information. Depending on the definition you choose, a transform may be lossy or lossless. For example, the projection of a three-dimensional scene into a two-dimensional image may be considered a transform. This is definitely a lossy transform—most of the depth information is lost in the projection, and we cannot accurately reverse the process. However, all of the transforms we consider in this chapter are lossless; the process can be reversed and will yield the original values—no information is lost.

This is an important point, and must be emphasized. We discuss mainly the *discrete cosine transform* (DCT). DCT is associated with lossy compression, but the transform is in itself lossless; with sufficient arithmetic precision, the transform can be reversed to any desired degree of accuracy. The value of DCT in compression is in the way in which it arranges information; it is another of the techniques that arrange data to be suitable for other compression tools.

One transform is already familiar to video engineers, the transform of color space from the tri-stimulus values of red, green, and blue (*RGB*) to the luminance plus color difference space represented by *Y, B-Y, R-Y,* or *YUV*. This transform is often represented by a set of three equations:

$$Y = \ \ 0.299 \times R + 0.587 \times G + 0.114 \times B$$
$$U = -0.299 \times R - 0.587 \times G + 0.886 \times B$$
$$V = \ \ 0.701 \times R - 0.587 \times G - 0.114 \times B$$

There is nothing wrong with this representation; each of the three equations is valid independently for a given set of *R, G, B* values. However, to use any one of the equations, we need values for each of *R, G,* and *B,* and that is all the information we need to apply all three equations to derive values for *Y, U,* and *V.* It is somewhat more meaningful to use matrix representation:

$$\begin{bmatrix} Y \\ U \\ V \end{bmatrix} = \begin{bmatrix} 0.299 & 0.587 & 0.114 \\ -0.299 & -0.587 & 0.886 \\ 0.701 & -0.587 & -0.114 \end{bmatrix} \cdot \begin{bmatrix} R \\ G \\ B \end{bmatrix}$$

This looks more "technical" and to some, perhaps unnecessarily complex. It's really just a different way of writing the three equations we started with, except for one thing. The matrix representation emphasizes that a *set* of *YUV* values is derived from a *set* of *RGB* values.

This transform is reversible; given a *set* of *YUV* values, we can calculate the equivalent *RGB*, because

$$B = Y + (B - Y) = Y + U$$
$$R = Y + (R - Y) = Y + V$$

and

$$G = \frac{1}{0.587}(Y - 0.299 \times R - 0.114 \times B)$$

$$= \frac{1}{0.587}(Y - 0.299[Y + V] - 0.114[Y + U])$$

$$= Y - 0.509 \times V - 0.194 \times U \quad \text{(approximately)}$$

or, if we can remember our matrix arithmetic or have access to mathematical software, we can invert the matrix and see that

$$\begin{bmatrix} R \\ G \\ B \end{bmatrix} = \begin{bmatrix} 0.299 & 0.587 & 0.114 \\ -0.299 & -0.587 & 0.886 \\ 0.701 & -0.587 & -0.114 \end{bmatrix}^{-1} \cdot \begin{bmatrix} Y \\ U \\ V \end{bmatrix}$$

$$= \begin{bmatrix} 1 & 1 & 0 \\ 1 & -0.19421 & -0.50937 \\ 1 & 0 & 1 \end{bmatrix} \cdot \begin{bmatrix} Y \\ U \\ V \end{bmatrix}$$

... again approximately. There is an important lesson here: the transform is, of course, perfectly reversible provided we have sufficient arithmetic precision. The values shown here, however, are not really accurate enough. If we use three significant figures for the input, these values give errors in the third significant figure of the output. Generally, we must work to a precision *at least* one decimal order greater than the accuracy we require.

We should consider another point before we leave the *RGB/YUV* transform. Note that there is no correspondence between one member of the *RGB* set and one member of the *YUV* set. We cannot say, for example, that *Y* corresponds to *G*, or that *U* corresponds to *R*. All we can say is that a *set* of *RGB* values can represent the same information, the same point in color space, as a *set* of *YUV* values. This is fairly obvious in this example, but important to remember when we are dealing with different transforms.

Time, Space, and Frequency Domains

The most successful compression systems (to date) are based on transforms that move between the time or space domain and the frequency or spatial frequency domain.

The fundamental building block for such transforms is the *Fourier theorem*, which states that any periodic function may be represented by the sum of appropriate amplitudes and phases of sine waves of one frequency and integer multiples of that frequency. This is quite easy to visualize, and Figure 5-1 shows an example.

Figure 5-1
Forming a complex waveform from components.

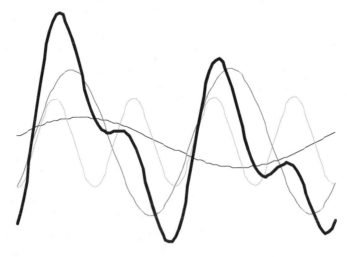

In its simplest form, the *Fourier transform* converts an infinitely long, but repetitive, function of time into an array of coefficients that express the amplitude and phase of each sine wave needed for the synthesis of the waveform. The array may be small if the waveform is simple. In fact, a pure sine wave is represented by a single Fourier coefficient pair—the amplitude and phase of that sine wave. If the signal is complex, the array of Fourier coefficients may be very large, but it is limited by the bandwidth of the signal; no Fourier terms are needed for frequencies outside the bandwidth of the signal. The more complex the waveform, and the higher its bandwidth, the larger the array of Fourier coefficients. At the theoretical extreme a signal with, for example, a perfect step (zero rise time, infinite bandwidth) requires an infinite array of Fourier coefficients to describe it.

Note the contrast between the two methods of describing a waveform. In the time domain we can easily determine what is happening (the value of the voltage or intensity) at any instant in time, but we can discover information about frequency only by examining all (or a large number) of time values together. Conversely, in the Fourier or frequency domain, we can easily see the frequency information, but we have no knowledge about events in time unless we examine all (or a large number) of the coefficients together. As with *RGB* and *YUV*, the *set* of frequency coefficients represents the same information as the *set* of time values, but no Fourier coefficient corresponds to an instant in time, or vice versa.

The simple Fourier transform has a number of limitations when we consider real-world problems:

1. It assumes that the time domain signal is infinite in extent.
2. It assumes continuous functions in time.
3. Up to now, we have considered only analysis of a one-dimensional function; images require two-dimensional analysis.
4. The coefficients generated are two-dimensional (amplitude plus phase), or complex (real plus imaginary), or amplitudes of both sine and cosine waves.

These issues have been resolved over the years, and today mathematics offers us tools derived from Fourier analysis that are very well suited to image analysis. One such tool is the *discrete cosine transform*. The mathematical derivation of the DCT is well beyond the scope of this book and the expertise of the author, but several references in the bibliography cover the subject in detail.

We will look at the properties of the DCT, but first we must clarify some issues of frequencies in time and space.

Frequency and Spatial Frequency

In the world of analog television we normally think of a signal as a variation of voltage with time; we also think of frequency as repetitions per unit time. This is quite valid in one dimension. We scan (or read out values from a *charge-coupled device* [CCD] array) horizontally, generating a signal whose voltage varies with time, and contains various frequencies. This is easy because, with conventional scanning, horizontally adjacent points in the picture become adjacent points in time in the signal. Vertically this model is less useful because vertical scanning is not a continuous function. Vertically adjacent points in the picture are separated in time, by one line's duration.

This asymmetry arises from the need to transmit a two-dimensional image with a one-dimensional signal (voltage varying with time). It is not a function of the image itself. A single image is unrelated to time; it is the scanning process that introduces the time factor by transmitting one picture element after another.

It is really important to separate this concept of time from that of time in a moving image. When we examine moving images we see that each two-dimensional image is a sample in time of the moving image. This is the concept of time that is important in imagery, not the time used to send data values sequentially.

To ensure that we do not confuse these issues, let's remove all concept of time from a static image. We will assume that sampling of the image is performed spatially by a CCD array, or similar device. Imagine that each element or pixel of the CCD is connected directly by a wire to the corresponding element of a display device. The process of conveying the information from pickup device to display device can be instantaneous or continuous; for a static image there is no difference. There is no defined scanning system; we can access the value of any pixel or combination of pixels according to our needs.

We have eliminated time from the static image, but frequency still has a meaning. Instead of voltage variations with time, we think of intensity variations with distance in the image plane. When we look at the analog signal derived from scanning we may see a time domain frequency of 3 MHz, or 3×10^6 cycles/s. This is just one possible way of conveying the information that the image contains an intensity variation of 160 cycles per picture width. This latter information is the property of the image.

Obviously a two-dimensional image can have intensity variations in any direction, but as with most two-dimensional phenomena we can analyze the plane by analyzing in two orthogonal directions, usually horizontally and vertically. Just as we must consider spatial frequencies in the image, we must analyze the image with sample densities measured spatially, for example, samples per millimeter or samples per picture height. Sampling in space must obey the same rules as sampling in time; the Nyquist limit is just as applicable, and the consequences of breaking the rules are the same in space as they are in time.

Fourier methods are still perfectly valid, but instead of functions of time and frequency, we deal with functions of distance (space) and spatial frequency.

Having eliminated time from our consideration of the static image, we can also clarify the relationship between pixels and samples. This is important because the two are often used interchangeably. Provided we understand the difference, this should not be a problem.

A pixel is the smallest element of an image that we can consider, or for which we can obtain a value. Physically, it may represent one cell of a CCD pickup device, *liquid crystal display* (LCD), or other display device. In the world of pickup tubes and *cathode-ray tube* (CRT) displays, it is set by the sampling rate applied to the scanned waveform, but should approximate the size of the beam spot.

In either case, a pixel clearly has physical dimensions. A certain number of pixels make up the width of the image; a certain number make up the height. A pixel is like a tile in a mosaic, except that in the world of image sampling all the tiles are the same size and shape, and are usually arranged in horizontal and vertical rows and columns.

A sample has only one dimension: the value of whatever it is we are sampling. In image sampling, that value is normally the intensity—either of luminance or of one of the color separations—of a single pixel.[1]

Just as a sample is considered to be an instant in time with no significant duration, a sample of an image should be considered to belong to a unique point in space. Having eliminated scanning and the time dimension from our analysis, our model should clearly be

---

[1] Different terminologies are possible. A sample of a color image could be described as three-dimensional in that it may represent a set of R, G, and B values, or Y, U, and V values. In compression processing we generally deal with one value at a time, for example, just the Y values, so I refer to the samples as single valued. The important point is that a sample has no *spatial* dimension.

symmetrical, thus it is best to assume that the point location of a sample is the geometric center of the corresponding pixel. This is also the best model for practical pickup and display devices.

This may sound very academic, but it is important to understand this relationship between pixels and samples because it is assumed in the DCT transform.

The Discrete Cosine Transform

As mentioned, a formal derivation of the DCT is beyond the scope of this book. However, an empirical understanding of the transform is essential to an appreciation of the compression schemes based on it. We examine several approaches to understanding DCT; all are valid—in fact, understanding how the different approaches are equivalent may be the best key to understanding the transform.

DCT can be defined for any rectangular array of pixels, but in image compression the basic block is generally an 8×8 array, or 64 pixels. Figure 5-2 shows such a block; the square tiles represent the pixels, and the dot at the center of each represents the location of the sample that carries the value of pixel intensity.

Figure 5-2
Block of pixels and the associated samples.

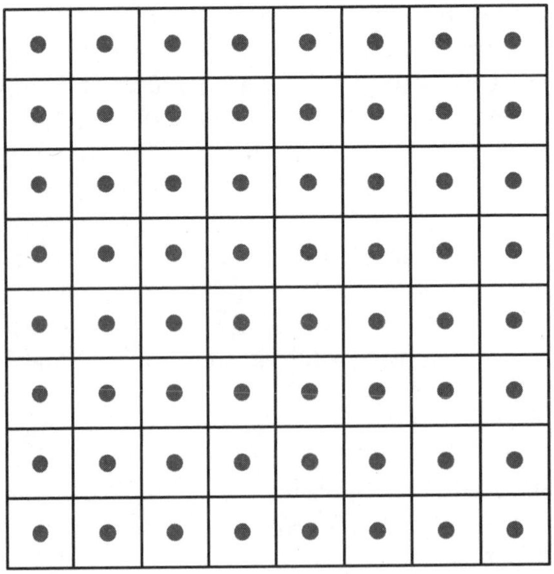

The DCT transform, as used in JPEG and similar compression schemes, is defined by the rather fearsome looking equation:

$$F(u,v) = \frac{1}{4} C_u C_v \sum_{x=0}^{7} \sum_{y=0}^{7} f(x,y) \cos\left(\frac{(2x+1)u\pi}{16}\right) \cos\left(\frac{(2y+1)v\pi}{16}\right)$$

where x and y are indices into an 8×8 array of samples, and u and v are indices into an 8×8 array of DCT coefficients. The C terms are defined by

$$C_u = \frac{1}{\sqrt{2}} \text{ for } u = 0, \quad C_u = 1 \quad \text{otherwise}$$

$$C_v = \frac{1}{\sqrt{2}} \text{ for } v = 0, \quad C_v = 1 \quad \text{otherwise}$$

The equation is not really as fearsome as it looks. Like most complex-looking mathematics, it is a shorthand representation. $F(u, v)$ means that the equation really represents 64 different equations, one for each value pair of u and v. So, for example, to evaluate the fourth coefficient in the third row (and remembering that the indices go from 0 to 7), we need to set $u = 3$ and $v = 2$. For these values of u and v, $C_u = C_v = 1$, so the equation for this one coefficient becomes

$$F(3,2) = \frac{1}{4} \sum_{x=0}^{7} \sum_{y=0}^{7} f(x,y) \cos\left(\frac{(2x+1)3\pi}{16}\right) \cos\left(\frac{(2y+1)2\pi}{16}\right)$$

Again, this is a shorthand; $f(x, y)$ means the value of the sample at x, y. Again, there are 64 of these, so for this one DCT coefficient, $F(3, 2)$, we have to evaluate the expression:

$$\frac{1}{4} f(x,y) \cos\left(\frac{(2x+1)3\pi}{16}\right) \cos\left(\frac{(2y+1)2\pi}{16}\right)$$

for each of the 64 values of $f(x, y)$, or for each pixel. The sigmas (Σ) mean that we have to sum the results of the 64 calculations. This is rather messy, and quite a lot of calculation, but nothing particularly complicated.

Note the special case that exists for $F(0,0)$, the top left DCT coefficient. Because the cosine of zero is one, the equation reduces to

$$F(0,0) = \frac{1}{8} \sum_{x=0}^{7} \sum_{y=0}^{7} f(x,y)$$

Apart from scaling, which varies with different implementations of DCT, the first DCT coefficient represents the average value of all 64 samples, or the average DC value of this part of the image.

Now let's examine some possible interpretations of the equation.

The Fourier Approach to DCT

DCT is most frequently characterized as a derivation of Fourier analysis, and we will examine this approach first. Conventional Fourier analysis applies to a continuous, periodic signal. In general the signals we wish to analyze are not periodic, and we will be dealing with samples rather than a continuous signal. Sampling presents no problem; the *discrete Fourier transform* (DFT) deals with a sampled signal, and is particularly valuable because there is an efficient approach to its implementation, known as the *fast Fourier transform* (FFT) which is very economical of computational resources.

It's worth taking a moment to look at the issue of Fourier components in the sampled domain. Fourier tells us that for a periodic waveform we need consider only the frequency of repetition (the sine wave with the same periodic time as the waveform) and integer multiples of that frequency, up to the bandwidth of the signal. In the sampled domain, we know the samples are meaningful only if the frequencies represented are lower than or equal to the Nyquist frequency, which is half the sampling rate. In other words, the highest frequency we need consider is that represented by two samples. Even if higher frequencies existed before sampling, they are now aliased down to below the Nyquist frequency.

Given this result, suppose we want to perform a Fourier transform on a periodic waveform where one period is represented by n samples. We have to find coefficients for the sine wave of period n samples, and each integer multiple of that frequency up to the frequency represented by two samples, *and no others.* As an example, if we have a sequence of eight samples, at a sampling frequency of *fs* we need to consider sine waves with frequencies:

$$\frac{f_s}{8}, \quad 2 \times \frac{f_s}{8}, \quad 3 \times \frac{f_s}{8}, \quad 4 \times \frac{f_s}{8}$$

The last frequency is half the sampling frequency—the Nyquist frequency—and we need go no higher. So our eight samples can be represented by four Fourier components, each requiring values for amplitude and phase, eight coefficients in all.

We can avoid, at least partially, the issue of periodicity by examining just a short segment of the signal, and pretending that it is repetitive. Let's take another example similar to Figure 5-1, but not repetitive, and consider just the first part of the signal, represented by eight samples as shown in Figure 5-3. In effect, we divide the time or space into eight equal regions and represent each region by a sample at its midpoint. (Typically, the sample value is the average value of the region.) If we perform a Fourier analysis on this segment, we can arrive at a set of coefficients representing the contributions of various frequencies. In fact, we assume that the signal is repetitive and looks like Figure 5-4.

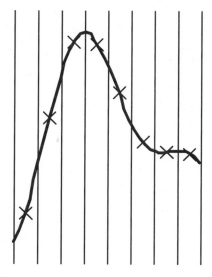

Figure 5-3
Samples of a waveform fragment.

This approach has two problems. Some part of the coefficients we generate will try to represent the sudden change in the signal after the eighth sample. This sudden change is a result of assuming that the signal is repetitive, and it is not part of the original signal. However, in performing a Fourier transform we have elected to represent the waveform

Figure 5-4
Combining
waveform fragments.

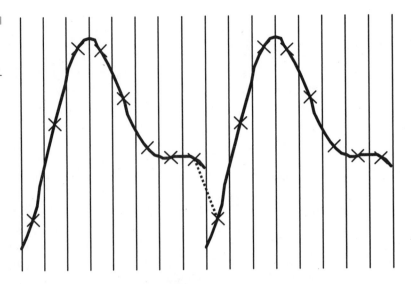

by a series of sine waves. These go on forever, and can only represent a repetitive waveform—the tradeoff is that the Fourier representation must include the transition from the last sample to a replication of the first. If the difference is significant, a considerable amount of the coefficient energy will be used in representing this nonexistent step.

We must also remember that each frequency in the transform is specified by both amplitude and phase. The transform coefficients are two-dimensional: amplitude and phase, real and imaginary, or cosine and sine (again, these are all equivalent). The easiest to consider at this time is cosine and sine. Any amplitude and phase of a given frequency may be obtained by summing appropriate amplitudes of sine and cosine waves of that frequency, one representing a 90° phase shift of the other. So, our signal segment of eight samples may be represented by four sine coefficients and four cosine coefficients.

Now, suppose that instead of analyzing a sequence of eight samples, we create a sequence of 16 obtained by mirroring the eight (see Figure 5-5). Because of the mirroring, the value of the last sample is the same as that of the first, thus no step is introduced by assuming this signal is periodic. Following the argument of the previous paragraphs we can conclude that we now need eight Fourier components to represent the new segment, so we might assume that we need 16 coefficients.

However, consider that

$$\cos(-x) = \cos(x)$$

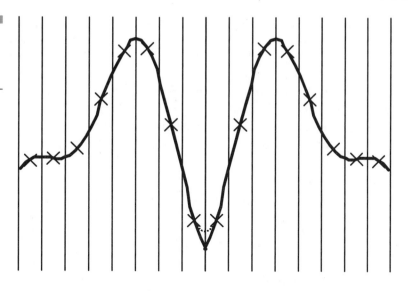

Figure 5-5
Mirroring a waveform fragment to create symmetry.

and

$$\sin(-x) = -\sin(x)$$

If we choose the center point as the origin we see that no sine components can be a part of the transform. All sine waves are asymmetric about their origin; our waveform is symmetric about its origin. This means that we can represent the 16 samples by eight cosine coefficients only. In effect, the mirroring of the samples causes us to use twice as many frequencies in the transform, but all the sine coefficients cancel out!

When we reverse the transform, the eight coefficients yield 16 mirrored samples, but of course we can discard the eight we do not need. This is the basis of DCT.

It turns out that DCT is a very efficient transform for photographic-type images—images sampled according to the Nyquist theorem. It is easier to handle than simple Fourier because we need only consider one type of coefficient (cosine). Also, because we have not introduced any large amplitude changes in our assumption of periodicity, DCT does not waste energy carrying information about the spurious high frequencies needed to represent these changes.

This explanation has addressed only a single dimension, but the principle is easy to expand to two dimensions. Instead of one set of eight samples, we analyze an image in space by taking a block of 64 samples. These can be considered as eight rows of eight samples, or as eight columns of eight samples, so that we can apply the analysis both horizontally and vertically. In fact, the DCT equation is separable; we can perform the horizontal analysis followed by the vertical analysis, or the other way around. The result is the same; in each case we transform a block of 64 intensity values into a block of 64 coefficients. As noted previously, the top left coefficient represents the DC level, or average intensity of the block. As we move to the right, the coefficients represent higher horizontal frequencies; as we move down, they represent higher vertical frequencies. The coefficient at the bottom right represents the combination of the highest possible horizontal and vertical frequencies.

In this approach, the DCT is visualized as a particular type of Fourier analysis, characterizing a block of pixel values by decomposing the block into a DC coefficient and 63 other coefficients representing the various spatial frequencies that may exist within that block of pixel values.

The lowest horizontal (or vertical) frequency that can be represented (apart from DC) is given by $u = 1$ (or $v = 1$), and is

$$\cos\left(\frac{(2x+1)u\pi}{16}\right) = \cos\left(\frac{1}{16}2\pi x + \frac{\pi}{16}\right)$$

This represents a cosine wave where the range $x = 0, ..., 7$ is one-half of a cycle. The wave is phase-shifted by $^\pi/_{16}$, or half of the pixel width. This places the maximum at the left-hand boundary of the block of pixels—one half pixel to the left of the first sample. Thinking back to the discussion of pixels and samples, this corresponds to the edge of the "tile" of pixels represented by the block of samples.

The highest is given by $u = 7$ (or $v = 7$) and is

$$\cos\left(\frac{(2x+1)u\pi}{16}\right) = \cos\left(\frac{7}{16}2\pi x + \frac{7\pi}{16}\right)$$

which has $3^1/_2$ full cycles across the block. The phase shift multiplies with the frequency, so that the effect at each frequency is the same— the maximum occurs at the edge of the pixel block, one-half of a pixel to the left of the first sample.

In essence, therefore, we can look at the DCT as a modified form of Fourier analysis. The inverse transform can similarly be considered a Fourier synthesis; both processes use cosine waves only.

DCT in Terms of Basis Functions

Many books on compression have covers representing the DCT basis functions, as shown in Figure 5-6. All transforms of this type have basis functions, and these can be regarded as sets of pixel values.

Figure 5-6
The DCT basis functions.

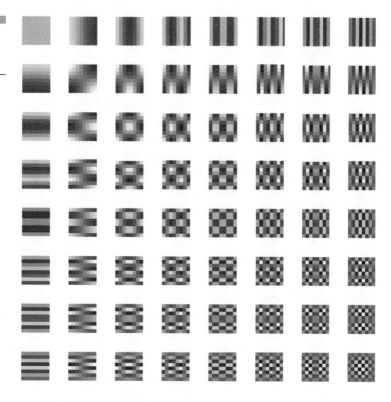

The inverse transform is expressed as

$$f(x, y) = \frac{1}{4}\sum_{u=0}^{7}\sum_{v=0}^{7} C_u C_v F(u, v) \cos\left(\frac{(2x+1)u\pi}{16}\right)\cos\left(\frac{(2y+1)v\pi}{16}\right)$$

where

$$C_u = \frac{1}{\sqrt{2}} \text{ for } u = 0, \quad C_u = 1 \text{ otherwise}$$

$$C_v = \frac{1}{\sqrt{2}} \text{ for } v = 0, \quad C_v = 1 \text{ otherwise}$$

As with the forward transform, the equation is evaluated 64 times, once for each pixel of the block; for example, $f(0,3)$ gives the value of the first pixel in the fourth row. Each evaluation uses every one of the 64 DCT coefficients, $F(u, v)$. Each DCT coefficient has an associated basis function; the basis function is the pixel pattern that results when that particular DCT coefficient is set to its maximum value, and all the other coefficients are set to zero.

As a simple example, if $F(0,0)$ is set to 256, and all the others are set to zero, every pixel in the block evaluates to 32. This is consistent with the Fourier view of DCT in which we saw that $F(0,0)$ was the DC coefficient, or a scaled average value of the block.

The basis functions represent 64 different arrays of pixel values. Given a set of DCT coefficients, the inverse transform operation consists of multiplying each basis function by its corresponding coefficient, the summing of the 64 sets of pixel values. The pixel values of the basis functions, and the different distribution of the values, mean that with an appropriate set of coefficients we can generate any pattern of pixel values.

If we take this inverse approach, the forward transform is the process of finding the set of 64 coefficients that, when multiplied by the 64 basis functions, yields the original 64 pixel values.

Figure 5-6 shows the full set of basis functions. The relationship to the Fourier approach is fairly obvious; as we would expect, as we move to the right, the basis functions represent increasing horizontal frequencies; as we move down, we see increasing vertical frequencies.

The exact relationship is easier to see if we look at a three-dimensional plot of the values of a basis function. Figure 5-7 shows the basis function values for $F(1, 0)$. If we extrapolate the calculation to include more points, we approach a continuous function, as shown in Figure 5-8. Now we can see the maximum and minimum of the curve occurring at the edges of the block, and see that the basis function values correspond to samples taken at the center of each pixel.

As a more complex example, Figure 5-9 shows the basis function values for $F(2, 5)$, and Figure 5-10 shows the corresponding quasi-continuous function.

Another way to visualize the use of the basis functions is to look at Figure 5-6 as if it were a paint box with textures instead of colors. To represent any array of pixel values (or, expressed another way, any pattern of 64 brightness values) imagine dipping a paintbrush into the basis-function paint box as many times as necessary. Given that both positive and negative coefficients are possible, it is not difficult to

Figure 5-7
The F(1,0) basis
function.

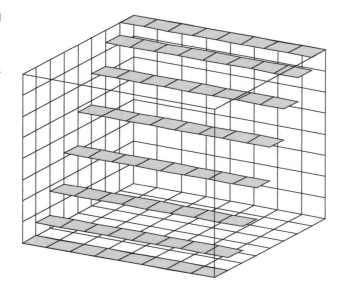

Figure 5-8
F(1,0) as a quasi-
continuous function.

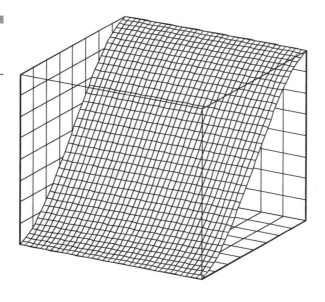

Figure 5-9
F(2,5) is difficult to
interpret in discrete
form.

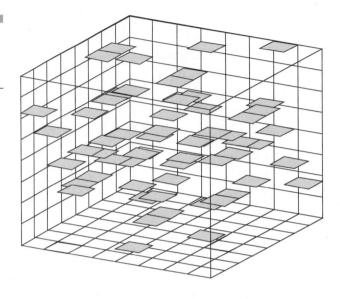

Figure 5-10
F(2,5) again, slightly
easier in quasi-
continuous form.

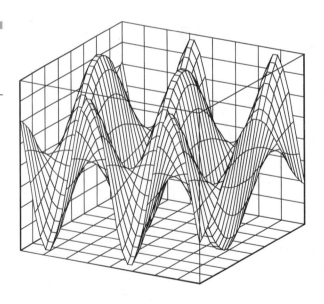

visualize that any required pattern can be built up by a sufficient number of dips into the paint box.

I particularly like this visualization because it is quite easy to imagine the result that makes DCT really useful for image compression. Most image segments can be represented to a good degree of accuracy by a relatively small number of DCT coefficients.

DCT as Axis Rotation

In Chapter 4 we looked at compression schemes using a predictor. Suppose we take an image and decide to code each pair of pixels. First we analyze the image and arrive at a large number of pixel pairs and their values x_1, x_2. If the values came from a memory-less source, there would be no correlation between the x_1 and x_2 values. However, a photographic image is similar to a Markov source; the most probable values for x_2 are those close to x_1. If we plotted the value pairs on axes representing x_1 and x_2, we might expect a distribution similar to that shown in Figure 5-11. The correlation is obvious to the eye when the values are plotted like this, but this does not help in coding the original values.

Figure 5-11
Array of sample pairs, x_1, x_2, where x_2 values are similar to x_1.

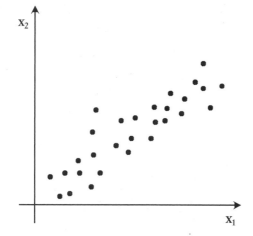

An obvious way to assist the coding is to use the first value of the pair as a predictor; we would then code x_1 and $(x_2 - x_1)$. Another approach that gives a similar answer is to rotate the axis frame so that one of the axes lies in the direction of greatest correlation, as in Figure 5-12. Now each pixel pair is represented by two new values, y_1 and y_2, where the majority of y_2 values are close to zero.

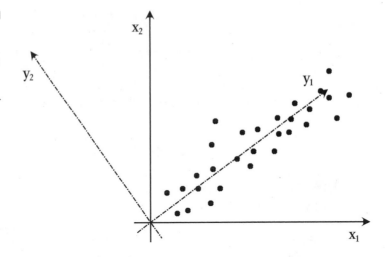

Figure 5-12
Rotation of the axes
makes the correlation
more useful; now all
y_2 values are small.

Now consider coding the samples in sets of three instead of two. Again we represent each triad by a single point on the plot, according to the values x_1, x_2, and x_3. Figures 5-13 and 5-14 show again that by rotating the axis frame appropriately we can minimize the energy of two directions. This is as far as we can go with a visual representation on paper, but it is not difficult to imagine that a similar approach is possible with larger sets of values.

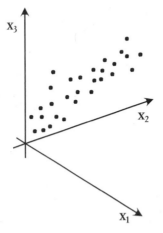

Figure 5-13
Correlated triplets.

Although we cannot visualize it, we can take an 8×8 block of pixels, and represent the block by a single point on a 64-dimensional plot. We can then rotate the axis frame in 64-D space (!) and obtain a new set of values that are much more efficient to code. If we choose the rotation

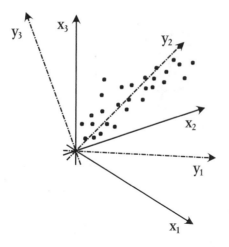

Figure 5-14
Correlation made
useful by axis
rotation: y_2 and y_3
have small values.

so that one of the axes passes through the point representing the average value of all 64 pixels, the new set of values is the set of DCT coefficients for the block.

DCT is one of a family of *distance preserving transforms*. When we rotate the axis frame, all the coordinate values change but the distance between points remains constant. It can also be shown that the total energy of the values remains constant (Rabbani, 1991). However, we have obviously changed the energy distribution; in fact we have concentrated most of the energy into a few coefficients.

This important property of the DCT makes it a valuable tool in compression systems. Note that DCT itself does not compress. The transform is accurately reversible given sufficient computational precision, but achieving this actually requires more precision to represent the transform coefficients than the original pixel values. As we will see in the next section, it is the energy concentration that permits us to reduce the number of bits used for coding while maintaining relatively good accuracy.

DCT Examples

Two simple examples help to show the process of the DCT. In Figure 5-15 the pixel block carries a section of a horizontal sine wave, and Figure 5-16 shows a horizontal raised cosine edge. As would be expected, these both transform to a set of coefficients that occupies only the first (top) row—a DC coefficient, plus horizontal-only AC coefficients.

Figure 5-15
This example shows a section of a horizontal sine wave; the transform shows horizontal coefficients only.

$$\begin{bmatrix} -17 & 18 & -154 & -1 & -25 & 0 & -7 & 0 \\ 0 & 0 & 0 & 0 & 0 & 0 & 0 & 0 \\ 0 & 0 & 0 & 0 & 0 & 0 & 0 & 0 \\ 0 & 0 & 0 & 0 & 0 & 0 & 0 & 0 \\ 0 & 0 & 0 & 0 & 0 & 0 & 0 & 0 \\ 0 & 0 & 0 & 0 & 0 & 0 & 0 & 0 \\ 0 & 0 & 0 & 0 & 0 & 0 & 0 & 0 \\ 0 & 0 & 0 & 0 & 0 & 0 & 0 & 0 \end{bmatrix}$$

Figure 5-16
A horizontal raised cosine edge also has a very simple transform again with horizontal coefficients only.

$$\begin{bmatrix} 418 & -540 & -345 & -136 & 0 & 41 & 25 & 4 \\ 0 & 0 & 0 & 0 & 0 & 0 & 0 & 0 \\ 0 & 0 & 0 & 0 & 0 & 0 & 0 & 0 \\ 0 & 0 & 0 & 0 & 0 & 0 & 0 & 0 \\ 0 & 0 & 0 & 0 & 0 & 0 & 0 & 0 \\ 0 & 0 & 0 & 0 & 0 & 0 & 0 & 0 \\ 0 & 0 & 0 & 0 & 0 & 0 & 0 & 0 \\ 0 & 0 & 0 & 0 & 0 & 0 & 0 & 0 \end{bmatrix}$$

DCT Failure

This section is really out of sequence, because I am going to show the effect of DCT combined with normalization and quantization, as it is used in JPEG and MPEG. However, I think the point is sufficiently important to make at this stage. DCT is useful for compression only if we can discard a large number of coefficients, and in general for blocks that obey the rules of continuous tone images this is true.

Figure 5-17 shows a complex block. There is a horizontal raised cosine edge at the top, and the same edge slightly displaced at the bottom. As we saw in Chapter 2, this generates vertical information, and in this case I have softened the vertical transition to the same rise time as the horizontal edge. Despite the complexity of this block, it survives the compression and reconstruction process well.

Figure 5-18 shows what might be thought to be a simpler block. The same edge is displaced, but to a greater degree, and there is no attempt

Figure 5-17
Even with quantization and normalization this displaced raised cosine (original at top) shows only minimal errors after processing (bottom).

to soften the resulting vertical transition. This generates not only vertical information, but also vertical alias. This breaks the rules around which the compression system was designed, and the figure shows the high degree of error that results. With the same degree of quantization, the block in Figure 5-18 suffers more than four times the mean square error than does the block in Figure 5-17.

The title of this section is rather unfair; it is not really DCT that has failed. We chose the DCT transform based on the assumptions made about the type of data to be transformed—bandwidth-limited images sampled according to Nyquist. The last example breaks these rules in that it contains a transition that creates aliasing. DCT is a powerful tool—provided we stick to the rules we assumed when the system was designed.

Figure 5-18
A large displacement
of the edge
generates a vertical
alias: with the same
level of quantization
as Figure 5-17, this
image shows much
greater distortion.

Quantization

Introduction

Quantization is the last of the individual tools we will examine before moving on to the use of the tools in combination to form complete compression schemes. We mentioned quantization in Chapter 2. We talked about the need to represent information with sufficient precision, but emphasized the importance of not transmitting unnecessary information.

Most of the tools we examined in earlier chapters do not by themselves reduce the amount of data. In fact, we saw that transforms and predictors can frequently generate more data. A predictor generates one additional bit per symbol; a transform like DCT generates three more bits per symbol. However, we have seen that some tools, such as predictors, arrange data in such a way that it is more suitable for the application of another tool, such as variable-length encoding. We will see shortly that the same is true when we consider quantization. We can quantize anything, but usually the greatest benefit is obtained if we use one or more manipulation tools to put the data into a form suitable for quantization.

Similarly, the tools we have examined up to now do not in themselves cause loss or distortion, provided enough precision is maintained. Even complex transforms such as DCT are fully reversible, given sufficient arithmetic precision.

When we first looked at quantization in Chapter 2, we were concerned about the degree of quantization necessary for optimal representation of the information. Too much precision just encodes noise and results in more data without more information. Now we are going to use quantization as a data reduction tool, and we will introduce loss. We will try to arrange the data so that only the least important information is lost by quantization, but when we use quantization *we discard information, and we will not get it back.* Quantization is irreversible. True, we will use a processing block called a *dequantizer*, but this is a very badly named process. A dequantizer reverses one of the side effects (scaling) of quantization, but does nothing to restore the precision that was discarded by the process.

Mean Square Error

Now that we have reached a point where we will definitely generate errors, it would be useful to have some means of quantifying them. It is possible to calculate the signal-to-noise ratio, generally defined as the ratio between peak signal and *root mean square* (RMS) error. Often we are concerned with minimizing the error rather than establishing an explicit relationship to the signal amplitude, and then we can omit a couple of steps. In describing the performance of quantizers and compression systems, it is common to refer to the *mean square error* (MSE). To determine the mean square error for any given set of values, the error of each value is squared (giving equal weight to positive and negative errors, but more weight to large errors). The squares are summed, and the result divided by the number of values to yield the MSE.

It is important to note that MSE is just one possible measure of error or distortion. It is a useful tool, provided its limitations are known and respected. MSE does not represent an accurate measure of subjective degradation of an image in the general case. The best that can be said is that for many images, and for many types of error, MSE correlates with subjective degradation as well or better than any other simple function.

The issues of error and distortion measurement are very complex. The only truly definitive measure of image quality is obtained by setting up subjective comparison tests using large numbers of observers, under controlled conditions. This process is time-consuming, very expensive, and quite impractical for real-time measurements or even day-to-day observations. There is a great deal of research going on to find objective measurement methods that can be used to rate image quality; such tools are badly needed for the evaluation and testing of compression systems. There has been some success. For example, Tektronix markets a "Picture Quality Analyzer" or "PQA" (based on technology developed jointly by Tektronix and Sarnoff Corporation) that provides a quality metric found to correlate well with subjective evaluation. This tool, and others on the market, have already contributed substantially to optimization of compression systems.

However, tools like the PQA are complex and slow, and there will continue to be many applications for a computationally simple test for error magnitude. Such tests may be used in real time to choose, for example, the best quantizer or the best predictor. MSE is a common test for these applications. In future chapters we will see the need for a similar test for block matching; MSE may also be used here, although *minimum absolute difference* (MAD) is also a common choice.

Types of Quantizer

There are two main divisions: scalar quantizers and vector quantizers. *Scalar quantizers* control the precision of expression of a single parameter, such as intensity. *Vector quantizers* approximate two or more parameters in a single value. Vector quantizers can be used alone as a method of compression and are very powerful for some applications. These are discussed later. For the present, we will look at types of scalar quantizers.

A scalar quantizer can be likened to a staircase function. The signal is constrained only to the values represented by the steps; "in between" values are not permitted. The greater the interval between the steps, the coarser the quantization.

Uniform Scalar Quantizer

The uniform scalar quantizer is the simplest form of quantizer, but often the most effective. It corresponds to a staircase with equal spacing of the steps, so that a continuum of input values is divided into a number of ranges of equal size. Figure 6-1 shows a range of input values from 0 to 255, divided by a uniform scalar quantizer into eight equal regions. There are 256 possible input values, but only eight possible output symbols. As the individual input values are processed by the quantizer, they are compared with decision values representing the steps of the staircase. Any input value between the first two decision levels is assigned the first output symbol, and so on.

Figure 6-1
Uniform scalar
quantizer.

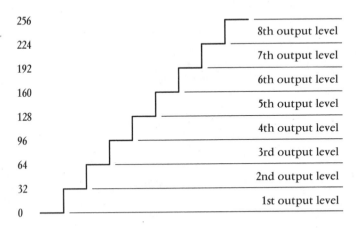

The other parameter that has to be considered is the reconstruction value. When we apply the quantizer shown in Figure 6-1, there are only eight possible output values, so we could represent each of these by a 3-bit code, or a value between 0 and 7, as shown in Table 6-1.

When we want to use the quantized data, it is no use using intensity values 0 to 7—these are all almost black. We need to get back to an approximation of the original range of input values, so for each of the quantized levels we need to choose a single value in the range 0 to 255. We could just add five zeros to each 3-bit symbol, but this would make all of the output values less than the input values; all the errors would be in the same direction.

It is reasonably intuitive that if the input values are evenly distributed, a reconstruction value at the center of each step is likely to give smaller overall errors and a lower MSE, as illustrated in Table 6-1. This is the reconstruction value chosen in the uniform scalar quantizer.

TABLE 6-1

Comparison of two sets of reconstruction values for the quantizer shown in Figure 6-1, and the resulting mean square error (MSE) values.

Input range	Output code	Reconstruction value		MSE	Reconstruction value		MSE
0–31	000	00000000	(0)		00010000	(16)	
32–63	001	00100000	(32)		00110000	(48)	
64–95	010	01000000	(64)		01010000	(80)	
96–127	011	01100000	(96)	325.5	01110000	(112)	85.5
128–159	100	10000000	(128)		10010000	(144)	
160–191	101	10100000	(160)		10110000	(176)	
192–223	110	11000000	(192)		11010000	(208)	
224–255	111	11100000	(224)		11110000	(240)	

Next we look at some rather more sophisticated quantizers. This is a useful exercise, but I warn you that at the end of the discussion, we will end up where we started; the uniform scalar quantizer is the best choice for many applications. However, the reasoning by which we reach this conclusion is quite interesting and instructive, so read on or skip to the next section—your choice!

Nonuniform Scalar Quantizers

In the discussion of the uniform scalar quantizer we assumed an even distribution of input values. Examination of the histograms in Chapter 4 shows that this is not a likely distribution for real images, and even less likely at the output of a predictor. In fact, the better the predictor the more values are zero or very close to zero. It is important to examine how this might affect the design of a quantizer.

Let's look at a hypothetical probability distribution shown in Figure 6-2. If we have only eight permissible output levels for the quantizer, we might speculate that a division of the input space similar to that shown might give a lower overall error than the even divisions of a uniform quantizer. Smaller errors are permitted in those areas where there are the most values (the decision values shown are just for example, not necessarily the best choices).

Figure 6-2
Possible nonuniform
quantizer.

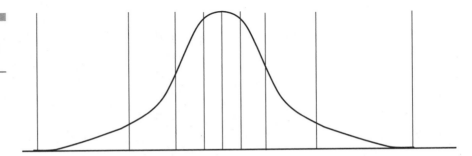

It seems reasonable to look for a quantizer that gives the lowest possible mean square error for a given number of quantizer output levels. One step that can help minimize the MSE is to make the errors smaller for large concentrations of values by placing narrower steps around these concentrations, and allowing larger step sizes (and errors) where there are relatively few inputs. This is the intuitive approach I took in drawing Figure 6-2.

The other part of the problem involves the choice of reconstruction values. Where there is an even distribution of input samples between two decision values, it is obviously reasonable to put the reconstruction value at the center of the range. But suppose the distribution is significantly skewed across the range, as in the outer ranges of Figure 6-2? We might expect a lower MSE if we shifted the reconstruction value toward the greatest concentration of samples.[1]

---

[1] But remember, MSE is not a good measure of subjective image degradation. The quantizer described achieves a lower average error value at the expense of a few larger errors. Which is worse? It depends on the image!

The Lloyd-Max Quantizer

In fact, the quantizer that gives the lowest MSE for a given number of output levels is constructed along these lines. It is known as the *Lloyd-Max quantizer*. Lim (1990) describes the derivation of this quantizer and gives tables of values suited to various probability distributions. We will not bother with the formal mathematics here, but a simple description is in order.

The Lloyd-Max quantizer places the reconstruction levels at the weighted mean value, or *centroid*, of each decision region (the region between two adjacent decision values). In other words, the position of the reconstruction value is chosen to minimize the total absolute errors within each decision region, taking into account the probability distribution of the input. The decision levels are placed halfway between adjacent reconstruction values. This sounds very easy until we realize that the derivation is iterative. In practice, a quantizer is calculated by making an inspired guess at the best decision values and calculating the reconstruction values for these decision regions. New decision values are then derived from the reconstruction values, and the process is repeated until there is no significant change.

Entropy-constrained Quantizers

The objective of the Lloyd-Max quantizer is to minimize the distortion when quantizing a source to a given number of levels. For example, if we want to quantize a source to 3 bits (8 levels), and if we know the probability distribution function for the source, the Lloyd-Max quantizer can be calculated to give us the lowest MSE for an eight-level quantizer applied to that source. This is good if we can meet all those conditions, but this still may not be the best answer if our real requirement is to transmit the source using the minimum number of bits.

The Lloyd-Max quantizer tends to distribute the decision zones to equalize the number of samples that fall into each zone. Thus, the probability of all output levels is similar and, in the case of an eight-level quantizer, the entropy of the quantizer output is close to 3 bits per sample. We can transmit this output with a fixed 3-bit code, but we are unlikely to gain significantly by using variable-length coding.

As an alternative, let's consider using a 16-level uniform quantizer. Now we have an output represented by 4 bits per sample, but if the source has nonuniform probability, some output levels will be more probable than others, and the entropy will be lower than 4 bits per sample, perhaps lower than 3 bits per sample!

So, if variable-length coding of the quantizer output is permitted, perhaps we should not be looking for the quantizer that provides the lowest distortion for a given number of output levels. If our goal is to transmit the image with as few bits as possible, the search should be for a quantizer that minimizes distortion for a given *entropy* measured at the quantizer output. This is the *entropy-constrained quantizer* (ECQ).

Now we come full circle. It has been shown theoretically that for a discrete memoryless source to be transmitted at high bit rates, the ECQ is the uniform scalar quantizer. Experimentally, it appears that the same quantizer is very close to the optimum ECQ for most bit rates for most probability distributions. In other words, for most applications in image compression, the simplest possible quantizer, followed by variable-length coding, produces the best results.

Vector Quantization

As described, a scalar quantizer acts upon a single, one-dimensional variable such as intensity. In the case of a color image expressed in *RGB* components, a scalar quantizer would operate on one of *R*, *G*, or *B*. In contrast, a vector quantizer operates on a multidimensional vector, a value represented by more than one coordinate.

The design of a vector quantizer starts with the definition of a set of *reconstruction vectors*; these are directly analogous to the reconstruction levels of the scalar quantizer, but each is a multidimensional value, such as a set of *RGB* components. Each reconstruction vector is represented by a single index for transmission (such as a color *number*), and together they are known as the *codebook* for the quantizer. Quantization is performed by comparing each (multidimensional) input with each of the chosen reconstruction vectors, and choosing the vector that represents the smallest error. MSE is the commonest measure of distortion, so for a three-dimensional input represented by components *x*, *y*, and *z*, the error measured when comparing an input to a reconstruction vector R is

$$MSE_R = \frac{(x_R - x_{in})^2 + (y_R - y_{in})^2 + (z_R - z_{in})^2}{3}$$

Vector quantization is a very powerful technique that can produce results very close to the theoretical limit, but it is computationally intensive. Codebook generation can be very complex and, unless shortcuts can be found, the testing of each input against every reconstruction vector is expensive. Rabbani (1991) gives an excellent introduction to these techniques for those who wish to delve deeper.

There is, however, an example of a simple vector quantizing system that is familiar to most people, and helps to illustrate the concepts. This is the *graphics interchange format* (GIF) developed by CompuServe, but placed in the public domain.

GIF takes an image represented by *RGB* values, each in the range of 0 to 255. This is the normal 24 bits per pixel representation of a color image in the computer world, capable of some 16 million colors. The GIF coder examines the image and picks a set of 256 colors, assigning a single 8-bit codeword to each. This codebook, consisting of 256 indexed sets of *RGB* values, is sent to the decoder. Each input pixel is then examined, and the nearest color of the 256 is chosen to represent it (the chosen color is the reconstruction vector). The pixel is then represented by the single 8-bit index for that color—an immediate compression of 3:1 before any other techniques are applied. In fact, the GIF coder goes on to use the LZW compression algorithm as described by Welch in 1984 (unfortunately this algorithm is patented).

This is not a trivial example; a full GIF encoder is quite complex. Examining an image using over 16 million colors and determining which 256 colors can represent it with minimum distortion is no simple task. Sometimes, however, the set of colors is determined by the known 256-color palette of the output device, so this most complex step is eliminated. Even then, there remains the task of determining which of the 256 colors best approximates each input *RGB*.[2]

---

[2] A full GIF encoder can be much more complex than that described. For example, GIF can use *color dithering* more closely to approximate a color in the original image. The coder generates pixels alternately between two or more colors from the codebook. When the reconstructed image is viewed from a normal distance, where the individual pixels are not visible, the eye sees a merged color that should closely match the color of the original image. Note that dithering introduces substantial high frequency energy into the reconstructed image. Such images are very unsuitable for DCT encoding.

Applications of Quantization

Quantization may be used in many ways; we will discuss three: direct quantization of the original image data, quantization in conjunction with predictive coding, and quantization in conjunction with transform coding, which will be discussed further in the next chapter, on JPEG.

Direct Image Quantization

In Chapter 2 we discussed quantization and how it defines the precision of the digital data. We can quantize image samples, or any other samples, to whatever precision is useful, but what are the practical limits? Obviously there is no point in quantizing to a greater precision than the accuracy of our processing electronics; if the same input does not generate the same output every time, we are generating data that just represents the internal noise of the equipment. Similarly, it is not helpful to generate data that represents only the noise present in the samples.

One issue that must be mentioned at this stage is that of *perceptual uniformity*. In the following discussion I assume that the original samples are taken from a signal that is perceptually uniform. In other words, a 1-percent change in value near black will be as perceptible or imperceptible as a 1-percent change near white. If we are dealing with gamma-corrected video (the usual case) this is sensibly true; note that it is *not* true for linear video. This is a big subject in itself, and beyond the scope of this book. (Charles Poynton has an excellent and very readable discussion in his book [Poynton, 1996].) Meanwhile, I continue to assume perceptual uniformity in the samples we are quantizing.

So, let's assume that we can quantize our perceptually uniform samples to the highest possible degree that still represents information rather than noise. This is clearly the optimum quantization in total information terms; is it optimum in terms of useful information? Obviously this depends on what we are going to do with it.

If we are going to process the data, to extract information from it, more information is better. Provided we can afford the additional precision of the processing hardware (or the additional time in software) it is generally better to use the highest precision available. Even if some accuracy is discarded after the processing, this approach gives better answers than introducing additional quantization noise before the processing. This is a purist's approach, and must be applied with common sense. There is little value in building a transform engine capable of accepting 12-bit precision if the output coefficients will always be quantized to 4 bits or less. Nevertheless, it is good practice to keep quantization noise the least significant factor until all processing is complete.

Once any processing is complete, the rules change. To be efficient, we must ensure that we do not deliver more information than can be used at the receiving point. If we are considering images to be viewed by the human eye, we must remember that the limit of human perception is about a 1-percent differential brightness over a contrast range of about 100 to 1. Ideally, we keep quantization noise below this, so 8 bits, 256 levels, represents a fairly good choice for delivering images that will fully exploit the capability of the human eye and provides a little quality overhead.

That choice is appropriate provided we will be delivering an image with a contrast ratio of 100:1. Actually, this is about the best we can do in a very dark movie theater; almost any other mechanism for image display will have lower contrast.

A good example is a halftone image in a book like this. A good quality white paper reflects just over 90 percent of the incident light. If you examine the large letters in a chapter title with a magnifying glass, you see the best black that the printer can achieve; it is very dark, but certainly not totally black. Typically, the best black in a half-tone image is somewhat less dense than this, perhaps 98 percent ink coverage. The situation is made worse by the fact that black ink reflects some light, particularly if it is at an angle to the page. If all of these factors combine to give only 3-percent reflectance off black regions of a half tone, our contrast ratio has dropped to about 30:1! Depending on paper, ink, printing, ambient lighting, etc., we may do better or worse than this, but it's certainly not 100:1.

So what quantization precision is appropriate for halftones? Provided we remember the need for perceptual uniformity, a reasonable rule of thumb is that the bare minimum number of levels is about equal to the expected contrast ratio; twice that number gives some margin of error and should be a safe number. This suggests that we may be able to get away with 5-bit quantization for halftones, and that 6-bit should be plenty.

I leave you to judge for yourself. Figure 6-3 shows four half tones of "Lena," reproduced with 3-, 4-, 5-, and 6-bit quantization (8, 16, 32,

(a) 3-bit (8 levels)

(b) 4-bit (16 levels)

(c) 5-bit (32 levels)

(d) 6-bit(64 levels)

Figure 6-3   Images of "Lena" at various quantization levels. Reproduced by special permission of Playboy magazine. Copyright © 1972, 2000 by Playboy.

and 64 levels of gray, respectively). I expect that if you study these closely under a variety of lighting conditions, the 3-bit quantization will always be very evident and objectionable. I do not expect you to be able to see the quantization at all on the 6-bit image. You may or may not be able to see the 5-bit quantization. If you hold the book so that the lighting is very oblique, there is more reflection from the black ink (lowering the contrast ratio still further), and even the 4-bit quantization may disappear.

So, if we are quantizing image data directly, we can choose the precision according to how we expect the image to be displayed. It is important to have enough precision, but if the display contrast ratio is low, we can save a good deal of data simply by adjusting the quantization to suit.

This analysis may also give us a clue as to the size of errors we may be able to tolerate if we use other coding schemes. With 6-bit quantization, the maximum error compared to original 8-bit data is ±2 and gives a mean square error of about 1.5. Remember these numbers in the next section, when we look at predictive coding again.

Quantization with Predictive Coding

In Chapter 4 we looked at predictive coding as a lossless technique. We saw that the image data could be spatially decorrelated by prediction, and that the resulting error values could be regarded as a DMS with quite low entropy. The benefits of this technique are real, and where lossless encoding is essential, it is one of the most powerful tools available.

Often, however, we are prepared to accept small errors in the reproduced image if we can save more bits. As a simple example, we saw that "Lena" quantized to 6 bits provided an acceptable halftone image; 5 bits may be acceptable for some applications. On the other hand, by predictively coding the same image in Chapter 4, we obtained a set of error values for transmission with an entropy of 4.3 bits/pixel—implying that with suitable coding we could transmit the image losslessly with less than 5 bits per pixel. Now let's consider quantizing the output of a predictive encoder.

For halftones, it appears that 16 levels provide a barely usable image, and 64 may be enough for a reasonable quality picture, so let's look at quantization that provides these numbers of levels. We will test uniform scalar quantizers with a step size of 16 and 4, and corresponding maximum errors for any pixel of ±8 and ±2 respectively.

The resulting images are shown in Figure 6-4. As we might expect, they are comparable in quality to the 16- and 64-level versions of Figure 6-3. For the 16-level image the MSE is about 22.5 in Figure 6-3 and 21.1 in Figure 6-4. The image in Figure 6-3b takes 4 bits per pixel to transmit, whereas the entropy of the quantized error signal for Figure 6-4a is 1.33. We should be able to transmit this version for about 1.5 bits per pixel.

Similarly for the 64-level images—both have an MSE of about 1.5. Figure 6-3d takes 6 bits per pixel; the entropy of the error signal for Figure 6-4b is only 2.7; we know that 3 bits per pixel will be plenty for this.

Chapter 4 showed that predictive coding is a powerful technique for lossless encoding. Combining predictive coding with quantization shows a considerable gain in efficiency for reasonable-quality images— certainly the combination is far more effective than quantization alone. However, we should be aware of two things before we move on.

First, note that all the visible errors in Figure 6-4 are in smooth areas of the picture where values change slowly and prediction errors are small. It is likely that we could improve the subjective quality if we could minimize the errors in these areas at the expense of allowing larger areas where values are changing rapidly and errors are less visible. These are also the areas where the prediction errors will be large, so it may be acceptable to permit larger quantization errors when encoding large error values. This is the behavior of a Lloyd-Max quantizer, thus this quantizer may be particularly useful for this application.

Second, a result of 3 bits per pixel for a good-quality monochrome image is not impressive when compared with more sophisticated compression systems. In the next chapter we will look at JPEG—a system that uses quantization with transforms as well as all the other techniques we have covered. At the end of that chapter, I will try to produce a subjectively similar quality with JPEG to compare the efficiencies of the two approaches.

(a) 16-level quantizer

(b) 64-level quantizer

JPEG

Introduction

JPEG refers to a committee that reported to three international standards organizations. The committee began work on a data compression standard for color images in 1982 as the Photographic Experts Group of the International Organization for Standardization (ISO). It was joined in 1986 by a study group from the International Telecommunications Union (then CCITT, now ITU-R), and became the Joint Photographic Experts Group (JPEG). In 1987 ISO and the International Electrotechnical Commission (IEC) formed a Joint Technical Committee for information technology, and JPEG continued to operate under this committee. Eventually the JPEG standard was published both as an ISO International Standard and as a CCITT Recommendation.

This committee represents an unprecedented degree of cooperation, not only among the organizations involved but also among the many experts from over 20 countries who created the standard. The members of the JPEG committee worked on the basis that all technical decisions had to be unanimous, not because of altruism, and certainly not because of a lack of conflicting ideas, but because no one knew how to resolve a dispute in a committee reporting to so many bodies!

We are fortunate that it was recognized at such an early stage that many industries and disciplines need compression of images. The JPEG process focused the attention of the world's experts so that instead of many overlapping and conflicting standards, we have a system that is well thought out and refined, applicable to many different uses, and rigorously tested.

At the outset it was decided that the core JPEG compression would address monochrome images, and that compression would be applied separately to the various components of a color image. Given an 8-bit/pixel monochrome image, the original targets were to provide "recognizable" pictures at 0.25 bits/pixel, "excellent" quality at 1.0 bits/pixel, and images "indistinguishable" from the original at 4.0 bits/pixel. Note that these targets did not represent particularly difficult tasks. We have seen in Chapter 4 that simple lossless techniques can provide compression of around 2:1 (4 bits/pixel), and lossless compression *must* yield results indistinguishable from the original. At the other extreme, we could replace each block of 8×8 pixels with just the average brightness. This represents a compression of 64:1 (0.125 bits/pixel) and the resulting image, although reduced in resolution by 8:1 in each direction, should certainly be "recognizable." As the work proceeded, the committee aggressively pushed performance requirements, and the targets were revised several times. Final testing was performed against targets of 0.083 (recognizable), 0.75 (excellent), and 2.25 (indistinguishable) bits/pixel, and an addition level of 0.25 bits/pixel was added for "useful" quality images.

Many compression schemes were tested, but eventually DCT was chosen as the core technology for JPEG. DCT provided the best pictures at low bit rates, and is relatively easy to implement with fast algorithms that can be built into hardware.

The definitive text on JPEG is Pennebaker and Mitchell (Pennebaker, 1993). This volume of more than 600 pages contains a wealth of information both technical and historical, as well as the complete text of the (draft) International Standard. The book also covers the many extensions to baseline JPEG. It is a "must have" for anyone implementing the standard.

In this book, we look at JPEG in less depth, but in enough detail to see how it uses the various tools we have already examined. We also play with the process and gain some understanding of JPEG's strengths and weaknesses. In particular we will look at what can go wrong in a JPEG environment, the things that "break" the process and can result in drastic loss of quality. This is not intended as a criticism of JPEG; used correctly it is a fine system. But it is important to remember that JPEG is intended for the compression of continuous tone images, *photographic* or *real-world* images as we have characterized them; images that do not conform to these constraints can suffer badly. Specifically, JPEG is not designed for binary (black/white) images, nor is it well suited to discontinuous images such as limited-color-palette images (e.g., GIF images).

Baseline JPEG

JPEG is an extensive standard offering a large number of options for both lossless and lossy compression. These options are summarized in a later section, but we are going to concentrate on the best-known implementation, known as *baseline JPEG*. This method uses DCT compression of 8×8 blocks of pixels, followed by quantization of the DCT coefficients and entropy encoding of the result.

Image Data Encoding

Baseline JPEG requires that an image be coded as 8-bit values. Compression is specified in terms of one value per pixel; color images are encoded and handled as (usually) three sets of data input. The JPEG standard is color-blind, in that there is no specification of how the color encoding is performed. The data sets may be *R*, *G*, and *B*, or some form of luminance plus color difference encoding, or any other coding appropriate to the application.

The most common coding used with JPEG is luminance (Y) plus C_B and C_R. C_B and C_R are derived from $(B - Y)$ and $(R - Y)$ respectively by scaling and shifting so that the full gamut for permissible *RGB* values can also be represented by an 8-bit value. Level 128 represents the zero color difference. Usually, the C_B and C_R signals are filtered to reduced spatial bandwidth. For 4:2:2 processing the horizontal bandwidth is halved; for 4:2:0 both horizontal and vertical bandwidths are halved.

JPEG Baseline Encoding and Decoding

First, the image is divided into its different components. Although the compressed data derived from the components may be interleaved for transmission (reducing the need for buffering at both ends), JPEG processes the various components quite independently. If the color components are bandwidth-reduced, the effect is merely that the coder operates on a smaller picture. For example, a Y, C_B, C_R 4:2:2 picture might be processed as one 720×480 image (luminance) and two 360×480 images (C_B and C_R), as shown in Figure 7-1. In the case of 4:2:0 processing, the C_B and C_R images would each be 360×240 pixels.

A simplified block diagram of both encoder and decoder for a single image is shown in Figure 7-2.

Each component (and from now on we shall consider only one) is divided into 8×8 blocks, the value 128 is subtracted from each pixel (see explanation below), and each block is transformed by DCT.

The next step is quantization. JPEG quantization is a sophisticated process that can effectively use a different quantizer for each coefficient in the 8×8 array. This process is one of the most significant contributors to the power of JPEG, and is described in detail below.

The quantized DC coefficient (top left) is separated from the AC coefficients, and the sequence of DC coefficients (one from each block of pixels) is predictively encoded, and then entropy-encoded.

Then comes a step we have not considered before. The quantized AC coefficients from a block are arranged and grouped to form *descriptors*. Like entropy encoding, this process is statistical, but at a higher level. The process uses the known characteristics of DCT and the quantiza-

Figure 7.1
Example
preprocessing of
color image for JPEG.

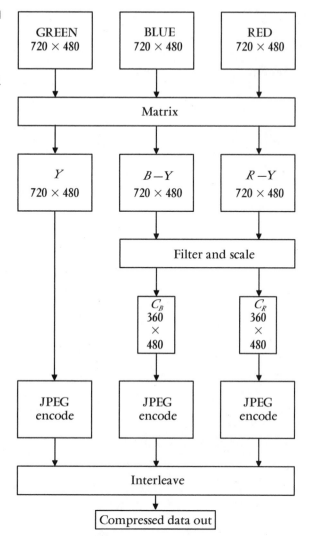

tion matrix to arrange and group the coefficients into units that are particularly suited to entropy encoding. These units are given descriptors (which are just like symbols, but we usually use this word to describe the output of the entropy encoder). Together the descriptors form the alphabet of a DMS, which is fed to the entropy encoder.

The data stream from the entropy encoder is combined with various markers to indicate data types and boundaries; these also serve to aid resynchronization at the receiver in the event of data errors.

At the receiver, the data stream is parsed to separate the different types of data. The entropy-encoded data from each pixel block is

Figure 7.2
JPEG encode and
decode.

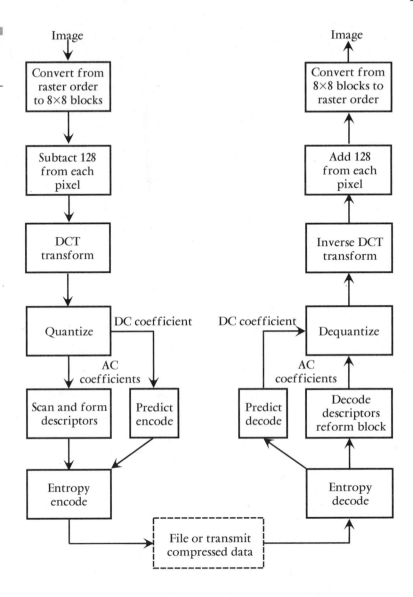

Figure 7.2
JPEG encode and decode.

decoded into descriptors which, together with the known descriptor definitions, allows the block of AC DCT coefficients to be reassembled. Separately, the stream of data representing the DC coefficients is decoded first by the entropy decoder, then by the predictive decoder, and the appropriate DC coefficient is associated with each group of AC coefficients.

The next step is known as *dequantization*. As previously noted, this is a poor name, as the process does not undo the effects of quantization. It does replace each quantized coefficient with the appropriate reconstruction value, taking into account the quantizer used for each individual coefficient.

All the coefficients of the block are now back in the correct range, the inverse DCT transform is applied, and 128 is added to each value to generate the block of received pixel values. These are reordered into the correct scan or file format, along with any other color components, and the process is complete.

From this summary it can be seen that JPEG is quite a complex process, but the individual steps are fairly simple. We'll take two blocks of 8x8 pixels all the way through the compression and decompression process, and examine what happens at each stage.

DCT Transform

The forward DCT equation as used by JPEG is

$$F(u,v) = \frac{1}{4} C_u C_v \sum_{x=0}^{7} \sum_{y=0}^{7} f(x,y) \cos\left(\frac{(2x+1)u\pi}{16}\right) \cos\left(\frac{(2y+1)v\pi}{16}\right)$$

where

$$C_u = \frac{1}{\sqrt{2}} \quad \text{for } u=0, \quad C_u=1 \quad \text{otherwise}$$

$$C_v = \frac{1}{\sqrt{2}} \quad \text{for } v=0, \quad C_v=1 \quad \text{otherwise}$$

As a reminder, to calculate each DCT coefficient (at location u, v) the equation sums contributions from each of the 64 pixels (at locations x, y). With pixel values, $f(x, y)$ in the normal range of 0 to 255, this equation yields AC coefficients in the range -1023 to $+1023$, and these may each be represented by an 11-bit signed integer. The DC coefficient, however, would be in the range 0 to 2040, and this *unsigned* 11-bit integer would require different handling in the hardware or software performing the DCT and subsequent operations. To avoid this discrepancy, the value 128 is subtracted from every pixel value prior to the DCT. This subtraction has no effect on the AC coefficients generated, but shifts the DC coefficient into the same range, so that all coefficients may be represented by 11-bit signed integers.

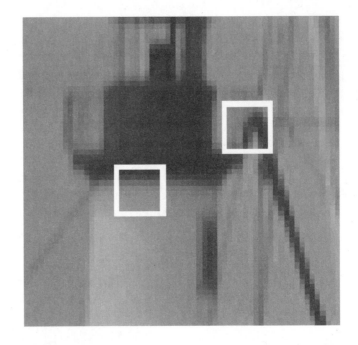

Figure 7-3
An enlarged section from the image "Boats" (top) showing the two pixel blocks used in the JPEG calculations. Below are the two 8x8 blocks, further enlarged.

Let's look at some real examples. I have chosen two blocks of pixels from the image "Boats," as shown in Figure 7-3. One is an "average" block—it has a vertical gradient and some texture. It is not exceptionally difficult, but many blocks in the image will be easier for the compression to handle. The other block has considerable detail horizontally and vertically, and high contrast. This block will stress the compression process and demonstrate the errors likely to arise in detailed areas of an image. Despite this difference in character, the entropy of the two sets of luminance data is similar, 5.04 and 5.59 respectively.

The two sets of values are

70	72	70	70	72	68	68	64
103	101	103	100	99	97	94	94
132	132	132	130	129	129	125	121
157	157	155	154	153	150	148	145
168	163	164	162	163	161	161	156
172	170	165	166	163	163	162	158
174	170	167	167	164	163	164	159
174	173	170	167	167	166	166	160

151	147	152	140	138	125	136	160
157	148	152	137	124	105	108	144
152	151	146	128	99	73	75	116
154	148	145	111	91	68	62	98
156	144	147	93	97	105	61	82
155	139	149	76	101	140	59	74
148	135	147	71	114	158	79	66
135	120	133	92	133	176	103	60

The first step is to subtract 128 from each value:

−58	−56	−58	−58	−56	−60	−60	−64
−25	−27	−25	−28	−29	−31	−34	−34
4	4	4	2	1	1	−3	−7
29	29	27	26	25	22	20	17
40	35	36	34	35	33	33	28
44	42	37	38	35	35	34	30
46	42	39	39	36	35	36	31
46	45	42	39	39	38	38	32

23	19	24	12	10	−3	8	32
29	20	24	9	−4	−23	−20	16
24	23	18	0	−29	−55	−53	−12
26	20	17	−17	−37	−60	−66	−30
28	16	19	−35	−31	−23	−67	−46
27	11	21	−52	−27	12	−69	−54
20	7	19	−57	−14	30	−49	−62
7	−8	5	−36	5	48	−25	−68

The DCT is then performed, yielding 64 coefficients, a single DC coefficient, and 63 AC coefficients.

86	2	−3	6	−2	2	−2	2
−247	−4	−5	−3	0	−3	1	1
−117	−1	1	−1	−1	−1	−1	0
−40	−2	2	1	2	1	−2	0
−7	−2	−1	1	0	−1	−1	2
−6	1	0	0	0	0	−2	−1
−4	−1	−1	1	−2	−1	−1	−1
−3	−3	−1	1	0	1	1	1

−63	157	22	11	−23	−37	68	11
62	−17	37	−90	64	15	−46	−2
66	−71	−25	37	−17	8	−2	4
16	−7	8	14	−14	−11	20	3
5	−12	−7	2	3	8	−12	2
−7	6	7	−7	−2	−3	−2	3
−3	−1	−4	0	1	1	0	−5
0	1	−1	0	−3	0	−2	0

Already we can see that DCT concentrates energy into the top left corner. This is particularly evident in the left example, the less-complex block. There are a few coefficients that represent the vertical gradient (particularly the three immediately below the DC coefficient), and a larger number of smaller coefficients representing minor corrections and texturing. In the other example, the concentration is also present but much less obvious; a large part of the energy of this block is in detail. We can also see the effect of the transform, and the difference in character between the two blocks, by looking at the entropy of the DCT coefficients, 1.42 and 2.72 respectively.

Quantization

The DCT coefficients shown above are already quantized to a small degree because of the 11-bit precision (all coefficient values have been rounded to the nearest integer). However, to achieve substantial compression we need to be much more aggressive. We have already seen in Chapter 6 that quantization of differential data can be much more effective (in terms of data saving versus quality loss) than quantization of the image data itself. The DCT coefficients of a continuous-tone image are very suitable for aggressive quantization, but here we move into the science of the *human visual system* (HVS). Study of the HVS reveals two important results for compression with DCT:

- When a block of pixels from a continuous-tone image is transformed, many of the DCT coefficients have values close to zero (in our examples there are 56 and 28 AC coefficients with magnitudes of 5 or less). The HVS is quite insensitive to errors in these coefficients; values close to zero may be set to zero with very little effect on image quality.
- The HVS is fairly sensitive to small errors in DC and low-frequency coefficients, but much less sensitive to amplitude errors in higher-frequency coefficients. We can afford to quantize such coefficients much more coarsely. (This not only results in using fewer bits to transmit each of the non-zero values, but also means that more coefficients can be treated as zero.)

In Chapter 5 we saw that each DCT coefficient can be viewed as the amplitude of the corresponding basis function. The sensitivity of the HVS to each basis function can be estimated by experiment, and this can lead to a possible quantization matrix. This experiment was performed (Lohscheller, 1984) and led to a pair of quantization matrices (one for luminance and one for color difference) that are included in Annex K of the JPEG standard. It is important to note that Annex K is informative and bears the title "Examples and Guidelines." The matrices are *not* part of the standard, nor does JPEG have any default tables; the tables used must be explicitly defined. That said, many JPEG applications do use these tables, or derivatives of them.

These tables represent quite aggressive quantization, and if JPEG compression is performed using these tables, the reconstructed images are likely to show significant degradation. Annex K says, "If these quantization values are divided by 2, the resulting reconstructed image is usually nearly indistinguishable from the source image."

16	11	10	16	24	40	51	61
12	12	14	19	26	58	60	55
14	13	16	24	40	57	69	56
14	17	22	29	51	87	80	62
18	22	37	56	68	109	103	77
24	35	55	64	81	104	113	92
49	64	78	87	103	121	120	101
72	92	95	98	112	100	103	99

17	18	24	47	99	99	99	99
18	21	26	66	99	99	99	99
24	26	56	99	99	99	99	99
47	66	99	99	99	99	99	99
99	99	99	99	99	99	99	99
99	99	99	99	99	99	99	99
99	99	99	99	99	99	99	99
99	99	99	99	99	99	99	99

The table for luminance (on the left) shows minima at (0,2) and (1,0). These represent minimum quantization, and so should correspond to maximum sensitivity of the HVS. Indeed, ignoring the horizontal-to-vertical discrepancy, these values do correspond quite closely to the maximum of the luminance contrast sensitivity curve for the HVS. The table for color difference (on the right) shows no minimum. This is reasonable, because the peak in the HVS contrast sensitivity curve for color difference functions is lower in frequency than the lowest basis function.

How are these matrices used? The values represent quantization intervals, thus the process is to divide each DCT coefficient by the corresponding value from the quantization matrix. To facilitate rounding and reconstruction, the magnitude of each coefficient is increased by half the quantization interval before the division; the result is then truncated to the nearest integer below. Reconstruction is then simply a matter of multiplying the quantized coefficient by the quantization interval. Figure 7-4 shows an example for a quantization interval of 4.

Figure 7-4
Quantization and
reconstruction in
JPEG.

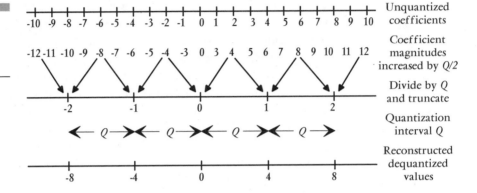

Now that we have been through all this, we can proceed with the processing of our sample blocks. We will use the luminance matrix specified above, even though it is rather too aggressive for high-quality work. At the end, we will discuss the effects of varying the quantization matrix.

We take each block of DCT coefficients, and to each coefficient $c_{u,v}$ we add half the corresponding quantization value $Q_{u,v}$, then divide by $Q_{u,v}$ and truncate to the nearest integer to derive the quantized coefficient $C_{u,v}$:

$$C_{u,v} = \text{floor}\left(\frac{c_{u,v} + \dfrac{Q_{u,v}}{2}}{Q_{u,v}}\right) = \text{floor}\left(\frac{c_{u,v}}{Q_{u,v}} + 0.5\right)$$

When we apply this function to our example coefficients, we obtain:

5	2	0	0	0	0	0	0
−21	0	0	0	0	0	0	0
−8	0	0	0	0	0	0	0
−3	0	0	0	0	0	0	0
0	0	0	0	0	0	0	0
0	0	0	0	0	0	0	0
0	0	0	0	0	0	0	0
0	0	0	0	0	0	0	0

−4	14	2	1	−1	−1	1	0
5	−1	3	−5	2	0	−1	0
5	−5	−2	2	0	0	0	0
1	0	0	0	0	0	0	0
0	−1	0	0	0	0	0	0
0	0	0	0	0	0	0	0
0	0	0	0	0	0	0	0
0	0	0	0	0	0	0	0

We can see that the number of nonzero coefficients is drastically reduced. Also, the quantization has reduced the number of values that must be transmitted. The entropy of our examples is reduced to 0.30 and 1.11.

Scanning, Descriptors, and Entropy Coding

We have seen that DCT concentrates energy into the "top left" coefficients. The quantization matrix emphasizes this trend, and we can see that the probability of a coefficient's being nonzero is much higher in the top left than at the bottom right. In fact, this probability almost exactly follows the diagonal zigzag pattern shown in Figure 7-5.

Figure 7-5
Zigzag scanning of
DCT coefficients.

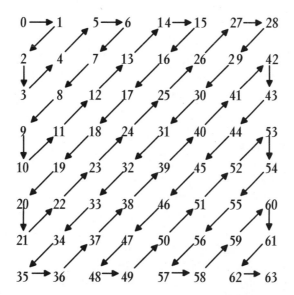

The output from the scanning process is entropy-encoded. JPEG specifies *two* possible entropy encoders; one based on Huffman coding and required for the baseline system is described below. JPEG extensions and lossless modes permit the use of the arithmetic coder (see Chapter 3) as an alternative.

When we scan the quantized coefficients of our first example, this pattern results in the following string of values:

(5) 2 −21 −8 0 0 0 0 0 −3 (all zeros)

and the second block gives:

$$(-4)\ 14\ 5\ 5\ -1\ 2\ 1\ 3\ -5\ 1\ 0\ 0\ -2\ -5\ -1\ -1\ 2\ 2\ 0\ -1\ 0\ 0\ 0\ 0\ 0\ 0\ 1\ 0\ -1$$
(all zeros)

The first value is the DC coefficient, and is shown in parentheses because it will be separated from the AC coefficients for entropy encoding.

The most striking characteristic of the sequences is that after a relatively small number of values, all remaining values are zero. We can take advantage of this immediately and directly by assigning a code symbol equivalent to "all zeros." This symbol is known as *end of block* (EOB). The zigzag scanning of the coefficients has given us some "free" compression. We are entitled to read out the values in any order we wish, provided this order is known at the decoder (thus putting coefficients back in the correct place in the block). Because of the distribution of the significant coefficients, zigzag scanning, as shown, almost always generates the longest runs of zeros, and in particular the longest run of zeros at the end of the scan.

Prior to the EOB condition, there are several occurrences of a string of zeros followed by one or more nonzero values. Again, this is characteristic of the value strings produced by the DCT transform and the zigzag scanning. JPEG uses a rather complex system of entropy encoding that takes advantage of this characteristic. The string is parsed (omitting the DC coefficient) into *descriptors,* and each descriptor includes a zero *run length* followed by a magnitude *category* for the value following the run of zeros. Each combination of run length and category is allocated a Huffman codeword. Because each category includes a number of possible values, this Huffman codeword is followed by a number of bits indicating which of the possible values in the category is the correct one.

This is nonintuitive and obscure, but perhaps the category list (Table 7-1) and a worked example will help to make it clearer.

For each category, additional bits equal in number to the category number are needed to define which member of the category is being coded. These bits are appended to the Huffman codeword specified for the run length/category pair. A short extract from an example Huffman table is given in Table 7-2.

Using this table we will parse the two examples, omitting the DC coefficients, which are processed separately.

TABLE 7-1

Categories and additional bits for AC coefficients

Category	Values included in category	Additional bits
1	$-1, 1$	0,1
2	$-3, -2, 2, 3$	00, 01, 10, 11
3	$-7, ..., -4, 4, ..., 7$	000, ..., 011, 100, ..., 111
4	$-15, ..., -8, 8, ..., 15$	0000, ..., 0111, 1000, ..., 1111
5	$-31, ..., -16, 16, ..., 31$	etc.
6	$-63, ..., -32, 32, ..., 63$	
7	$-127, ..., -64, 64, ..., 127$	
8	$-255, ..., -128, 128, ..., 255$	
9	$-511, ..., -256, 256, ..., 511$	
10	$-1024, ..., -512, 512, ..., 1023$	

For the first example, the AC coefficients are encoded to a total of 40 bits:

String	2	−21	−8	0 0 0 0 0 −3	All zeros
R/C	0/2	0/5	0/4	5/2	EOB
Codeword	01	11010	1011	111111110111	1010
Add. bits	10	01011	0111	00	

while the second block's AC coefficients are encoded to 94 bits:

String	14	5	5	−1	2	1	3	−5	1	0 0 −2
R/C	0/4	0/3	0/3	0/1	0/2	0/1	0/2	0/3	0/1	2/2
Codeword	1011	100	100	00	01	00	01	100	00	11111001
Add. bits	1000	111	111	0	10	1	10	010	1	10

String	−5	−1	−1	2	2	0 -1	0 0 0 0 0 0 0 1	0 −1	All zeros
R/C	0/3	0/1	0/1	0/2	0/2	1/1	7/1	1/1	EOB
Codeword	100	00	00	01	01	1100	11111010	00	1010
Add. bits	010	0	0	10	10	0	1	1	

TABLE 7-2

Partial table of
Huffman codes
for luminance AC
coefficients

Run/category	Code length	Codeword
EOB	4	1010
0/1	2	00
0/2	2	01
0/3	3	100
0/4	4	1011
0/5	5	11010
...	...	...
1/1	4	1100
1/2	5	11011
...	...	...
2/1	5	11100
2/2	8	11111001
2/3	10	1111110111
...	...	...
3/1	6	111010
3/2	9	111110111
3/3	12	111111110101
...	...	...
4/1	6	111011
4/2	10	1111111000
4/3	16	1111111110010110
...	...	...
5/1	7	1111010
5/2	11	11111110111
5/3	16	1111111110011110
...	...	...
6/1	7	1111011
6/2	12	111111110110
...	...	...
7/1	8	11111010
7/2	12	111111110111
...	...	...

A study of these results shows that the coding scheme is really very clever. After the DCT and quantization there are usually only a very few AC coefficients with large values. Most of the values are zero, and this fact is utilized both in the use of EOB, and by coding run lengths of zeros as shown. Most of the nonzero coefficients are small, and it is clear that this coding scheme is very efficient with small values.

In Chapter 4 we saw the power of the modified Huffman code. The JPEG coding just described is a form of multi-level modified Huffman code. Instead of dividing the alphabet into two zones, one with coded values and the other for "everything else," each category is treated as a zone with its own defined set of possible values, and its own number of additional bits to determine which member of the category was transmitted. Note that each category has a missing "core" corresponding to the sum of all the lower categories, and the number of values remaining is always a power of two. Not a single bit is wasted!

The DC coefficients (one from each DCT block) are treated in a similar manner, but with an extra step. Because the DC coefficient represents the average intensity of the block, there is still significant correlation between adjacent blocks. These values are predictively coded as described in Chapter 4, using the previous block's quantized DC coefficient as the predictor. This produces a single error value for each block and, as we saw in Chapter 4, there is a very high probability of this error, being small. The errors are coded using categories—the same technique and table used for the AC coefficients but with the addition of a category "0". A different Huffman table is used (Tables 7-3 and 7-4).

We did not examine data from previous blocks, so we cannot make the prediction and determine exactly how many bits would be needed for the DC components of the two example blocks. However, on the basis of the results in Chapter 4, it is safe to assume an average value of not more than 4 bits for each DC component. If we take this value, the compressed data for the first example block is 40 + 4, or 44 bits total. The original data for the block was 64 pixels, each with an 8-bit value. The compressed version has about 0.69 bit/pixel, corresponding to a compression ratio of 11.6 to 1. The complex example fared less well; it used 98 bits, or 1.53 bits/pixel, for a compression ratio of 5.2 to 1.

Bear in mind that many blocks in a real image are much less complex than even our "simple" example. Such blocks will generate even fewer AC coefficients, and many will be coded with a DC coefficient only for a total of (say) 4 bits, or a compression ration of 128-to-1 (predicted).

TABLE 7-3

Categories and additional fits for DC coefficients

Category	Values included in category	Additional bits
0	0	—
1	$-1, 1$	0,1
2	$-3, -2, 2, 3$	00, 01, 10, 11
3	$-7, \dots, -4, 4, \dots, 7$	000, ... , 011, 100, ... , 111
4	$-15, \dots, -8, 8, \dots, 15$	0000, ... , 0111, 1000, ... , 1111
5	$-31, \dots, -16, 16, \dots, 31$	etc.
6	$-63, \dots, -32, 32, \dots, 63$	
7	$-127, \dots, -64, 64, \dots, 127$	
8	$-255, \dots, -128, 128, \dots, 255$	
9	$-511, \dots, -256, 256, \dots, 511$	
10	$-1024, \dots, -512, 512, \dots, 1023$	

TABLE 7-4

Example of Huffman codes for luminance DC coefficients

Category	Code length	Codeword
0	2	00
1	3	010
2	3	011
3	3	101
4	3	101
5	3	110
6	4	1110
7	5	11110
8	6	111110
9	7	1111110
10	8	11111110

Assembling the Bit Stream

We have covered most of the steps used in JPEG to form the compressed image data. However, to make this data useful, there are a few additional steps. We must identify the beginning, end, and shape of the image—how many blocks horizontally and how many vertically. We must identify which component of an image is which (Y, C_B, C_R, for example). Quantization and Huffman tables must be specified unless previously sent separately. Also, the use of variable-length coding means that it is difficult to resynchronize after any data corruption. "Restart" markers are necessary from time to time to ensure that resynchronization is possible.

JPEG provides a complete system of markers, always identified by two consecutive bytes with value FF (hex)—in other words, bytes where all 8 bits have the value 1. The byte following the marker defines what type of marker it is.

Now that we have made the point that these mechanisms are necessary, and that they are provided, further discussion of the bit stream syntax is outside the realm of this book. If you want to build a JPEG system, read Pennebaker and Mitchell.

Parsing the Received Bit Stream

The first step of decoding is to separate the received bits into symbols that we can decode. The previous section briefly described the use of markers for synchronization and for identifying the different types of data, which permits us to progress to the point where we know we have the start of compressed image data.

The process is the same for both DC and AC coefficients. The incoming bit stream is examined and compared to the table of Huffman codewords. Huffman codes are *prefix condition codes* (see Chapter 3), so this is unambiguous—as soon as we find a match we extract the codeword. In the case of DC differentials, the codeword specifies a category; for AC coefficients it specifies both a zero run length and a category. In either case, the category specifies how many additional bits must be extracted from the bit stream to complete the decoding of this descriptor. The combination of Huffman codeword plus additional bits is then passed on for decoding and the parser looks for the next Huffman code.

Recovering the Quantized Coefficients

The DC coefficients were predictively coded, so decoding the Huffman code and additional bits gives a correction value that must be applied to the DC value of the previous block (the predictor).

For the AC coefficients, let's follow the process for the simpler block. After removal of the bits representing the DC differential, the remaining bit stream is

011011010010111011011111111111110111001010

To parse this, we first search for a codeword that appears in the Huffman table, starting with 2 bits and adding 1 bit until we get a match. This time we need only look at the first two bits; the codeword 01 appears in the table. It corresponds to category 2, so we must take an additional 2 bits from the bit stream to determine which member of category 2 was coded.

01	Run length = 0; category = 2
10	Value is 2
11010010111011011111111110111001010 remains	

Now we look at the bit stream for another Huffman codeword: 11 does not appear in the table, nor does 110 or 1101. Then we get a hit: 11010 is a codeword. It corresponds to a run length of zero, and category 5, so we must take the next 5 bits to determine the value within category 5.

11010	Run length = 0; category = 5
01011	Value is −21
10110111111111110111001010 remains	

Continuing like this,

1011	Run length = 0; category = 4
0111	Value is −8
111111110111	Run length = 7; category = 2
00	Value is −3
1010	EOB

When we find EOB we know that all the remaining coefficients are zero. Thus, we have a sequence of AC coefficients. When we add back the DC coefficient and place in the matrix according to the zigzag scan, we get back to the coefficients we saw after the quantization step. (Both examples are shown again.)

5	2	0	0	0	0	0	0
-21	0	0	0	0	0	0	0
-8	0	0	0	0	0	0	0
-3	0	0	0	0	0	0	0
0	0	0	0	0	0	0	0
0	0	0	0	0	0	0	0
0	0	0	0	0	0	0	0
0	0	0	0	0	0	0	0

-4	14	2	1	-1	-1	1	0
5	-1	3	-5	2	0	-1	0
5	-5	-2	2	0	0	0	0
1	0	0	0	0	0	0	0
0	-1	0	0	0	0	0	0
0	0	0	0	0	0	0	0
0	0	0	0	0	0	0	0
0	0	0	0	0	0	0	0

Dequantization

Quantization is, of course, the irreversible step. However, we do need to get each coefficient back in the correct range, and we do this by replacing every quantized value with the appropriate *reconstruction* level. Because we added half the quantization step to each value before dividing (Figure 7-2), all we need to do now is multiply each value by its quantization step, in other words by the corresponding value in the quantization matrix. This gives the following values:

80	22	0	0	0	0	0	0
-252	0	0	0	0	0	0	0
-112	0	0	0	0	0	0	0
-42	0	0	0	0	0	0	0
0	0	0	0	0	0	0	0
0	0	0	0	0	0	0	0
0	0	0	0	0	0	0	0
0	0	0	0	0	0	0	0

-64	154	20	16	-24	-40	51	0
60	-12	42	-95	52	0	-60	0
70	-65	-32	48	0	0	0	0
14	0	0	0	0	0	0	0
0	-22	0	0	0	0	0	0
0	0	0	0	0	0	0	0
0	0	0	0	0	0	0	0
0	0	0	0	0	0	0	0

Inverse DCT

The dequantized coefficients are now used as inputs to the inverse DCT.

$$f(x,y) = \frac{1}{4}\sum_{u=0}^{7}\sum_{v=0}^{7} C_u C_v F(u,v) \cos\left(\frac{(2x+1)u\pi}{16}\right)\cos\left(\frac{(2y+1)v\pi}{16}\right)$$

where

$$C_u = \frac{1}{\sqrt{2}} \text{ for } u=0, \quad C_u=1 \text{ otherwise}$$

$$C_v = \frac{1}{\sqrt{2}} \text{ for } v=0, \quad C_v=1 \text{ otherwise}$$

−54	−55	−56	−57	−59	−60	−61	−62
−29	−30	−31	−32	−34	−35	−36	−37
4	3	2	1	−1	−2	−3	−4
28	27	26	24	23	22	20	20
37	36	35	34	32	31	30	29
9	38	37	36	34	33	32	31
42	41	40	39	37	36	35	34
45	45	44	42	41	39	38	38

19	31	13	19	18	−9	4	32
25	35	17	8	−3	−29	−21	9
23	32	18	−10	−28	−49	−47	−14
17	23	15	−26	−43	−49	−59	−28
19	17	16	−34	−41	−32	−63	−43
27	13	17	−36	−27	−1	−61	−61
24	0	9	−38	−7	35	−48	−67
3	−15	−1	−41	7	62	−33	−63

The values above are the rounded outputs of the inverse DCT. Finally we add 128 to each pixel value to reach the pixel values for the reconstructed image.

74	73	72	71	69	68	67	66
99	98	97	96	94	93	92	91
132	131	130	129	127	126	125	124
156	155	154	152	151	150	148	148
165	164	163	162	160	159	158	157
167	166	165	164	162	161	160	159
170	169	168	167	165	164	163	162
173	173	172	170	169	167	166	166

147	159	141	147	146	119	132	160
153	163	145	136	125	99	107	137
151	160	146	118	100	79	81	114
145	151	143	102	85	79	69	100
147	145	144	94	87	96	65	85
155	141	145	92	101	127	67	67
152	128	137	90	121	163	80	61
141	113	127	87	135	190	95	65

Comparison

How well did it work? We have taken these blocks through the complete JPEG compression and decompression process; is the output similar to the input? Let's calculate the differences pixel by pixel:

4	1	2	1	−3	0	−1	2
−4	−3	−6	−4	−5	−4	−2	−3
0	−1	−2	−1	−2	−3	0	3
−1	−2	−1	−2	−2	0	0	3
−3	1	−1	0	−3	−2	−3	1
−5	−4	0	−2	−1	−2	−2	1
−4	−1	1	0	1	1	−1	3
−1	0	2	3	2	1	0	6

−4	12	−11	7	8	−6	−4	0
−4	15	−7	−1	1	−6	−1	−7
−1	9	0	−10	1	6	6	−2
−9	3	−2	−9	−6	11	7	2
−9	1	−3	1	−10	−9	4	3
0	2	−4	16	0	−13	8	−7
4	−7	−10	19	7	5	1	−5
6	−7	−6	−5	2	14	−8	5

It's difficult to know quite what the numbers mean. Obviously there are many small differences from the input, and there are a few quite large differences in the right-hand block.

When we look at the numbers, the MSE for the left-hand block is 6.125, corresponding to a signal-to-noise ratio of 40.26 dB. The more complex block obviously suffered more; the numbers are 52.89 and 30.9 dB.

More important, what do the blocks look like? Figure 7-6 shows before and after for each block. There clearly are visible differences; much of the texture has been lost from the simple block, and there are some significant contrast changes on the edges in the complex block. Equally obviously, the major elements of the image have been retained. We must look at some complete images to form a useful judgment.

Examples of Baseline JPEG

Figures 7-7 to 7-10 show examples of fairly aggressive JPEG compression on two images, "Boats" and "Lena." The compression ratio is about 22:1 for both images, corresponding to about 0.36 bits/pixel.

Figure 7-6
Before (left) and after
(right) examples of
JPEG processing.
These are the two
examples worked in
the test.

On the one hand, it is easy to see the results of compression. There is loss of detail particularly in "Boats," and very evident blocking artifacts in the sky of "Boats," and in many areas of "Lena." I chose a large compression ratio to make the effects easily visible in this book. Nevertheless, the results are not all that bad. Hold the book at arm's length and the compression errors are not very evident.

In the enlarged views, the mechanism of JPEG can be seen clearly. The small blocks are pixels, and you can clearly see the grouping into blocks of 8×8. It is easy to see that some blocks were coded with the DC coefficient only—they appear completely flat. In other blocks you can see patterns clearly resembling the low-order basis functions.

JPEG Extensions

In this chapter I have described the fundamentals of baseline JPEG. The standard provides for many extensions, some of which are briefly described below.

Baseline JPEG specifies a precision of 8 bits, but this is inadequate for applications such as medical imaging. One of the JPEG extensions permits 12-bit precision. This implementation, of course, requires considerably greater computational resources, particularly for the DCT and IDCT.

Another JPEG extension permits adaptive quantizing. A 5-bit scale factor may be applied to the quantization matrix on a block-by-block basis. This permits the encoder to use less quantization (resulting in more bits) for complex parts of the picture, and more quantization (fewer bits) where the image is simple. For any given image, application of this technique results in better quality for the same number of bits. It also permits control of the number of bits used by each image, which provides a mechanism for rate control in motion JPEG (see below). MPEG (see Chapters 9 and 10) adopted this approach, and adaptively coded JPEG images can be transcoded to MPEG intra frames.

A useful set of JPEG extensions is often seen in downloading JPEG images from the Internet. The image can be made to build up, either spatially or by resolution, so that a low-grade image may be seen before downloading is complete.

Motion JPEG

Motion JPEG is not really covered by the JPEG standard, but it provided a powerful tool for compressing motion sequences prior to the arrival of MPEG.

The concept is simple—each frame of an image sequence is coded as a JPEG image (usually with adaptive encoding to provide a fixed number of bits per frame) and the frames are transmitted or stored sequentially. This is the basis of most nonlinear editors and most compressed videodisc stores today.

Unfortunately, because motion JPEG is not properly standardized, virtually every implementation is proprietary and different. Usually image exchange must be performed by decoding in one equipment and recoding in another. This is inefficient, but no alternative yet exists.

New applications are unlikely to use motion JPEG; MPEG can be used in a motion JPEG-like mode and provides an environment where the adaptation and bit-rate control are standardized, permitting interchange of compressed images. Nevertheless, some form of motion JPEG is used in most nonlinear editors and, in general, backwards compatibility is more important than compressed interchange.

Figure 7-7
Before (top) and after
(bottom) JPEG
compression. The
lower image has
been compressed by
about 22:1.

Figure 7-8   Enlarged detail from the images of Figure 7-7.

Figure 7-9
Again, the lower image has been compressed by about 22:1. This image has many smooth gradients, and suffers badly from blocking artifacts. Reproduced by special permission of Playboy magazine. Copyright © 1972, 2000 by Playboy.

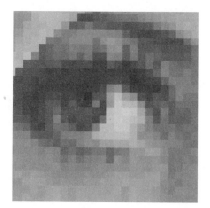

Figure 7-10   Enlarged detail from the images of Figure 7-9. Reproduced by special permission of Playboy magazine. Copyright © 1972, 2000 by Playboy.

Motion Compensation

Introduction

In previous chapters we explored the concept of spatial correlation within a continuous-tone image, and we have seen that we can exploit this correlation to compress the image data. In Chapter 7 we saw that the JPEG system for compressing static images could be applied to a sequence of images, compressing each individually. This technique is known as *motion JPEG*.

In motion JPEG, it is clear that the compression efficiency is a function of each individual image, and of that image only. Each image in the sequence is compressed according to the same rules; it makes no difference whether the images are totally unrelated, similar in content, or even identical. Motion JPEG takes no advantage of any correlation between successive images.

Temporal Redundancy

It is easy to see, however, that there should be considerable benefit available here. A sequence of images representing a moving scene is a set of temporal samples. In a typical scene there will be a great deal of similarity between nearby images of the same sequence. If you hold a strip of motion picture film up to the light, the repetitive nature of the images is evident. This means we should have some success predicting one from the other. As in the spatial domain, we should be able to use this property to separate and eliminate redundant data.

In Chapters 2 to 6 we looked at the various operations that would form our tool set for compression of continuous-tone static images. In Chapter 7 we saw how all of these were brought together to form the JPEG system. Now we add one more major tool, *motion compensation*. This allows us to go on to Chapters 9 and 10, where we study the MPEG-1 and MPEG-2 coding systems, designed for motion imaging.

Motion Aliasing

It might reasonably be assumed that we could extend the tools we have been working with from two dimensions to three. Certainly the discrete cosine transform can be defined for three dimensions, just as it can for one or two. In fact, this approach works for images that are soft and have only very slow motion. Unfortunately, this is not the situation we face in normal work. We saw in Chapter 2 that if the position of a horizontal edge changes between vertical samples, vertical information is created. Too much change per sample, and vertical aliasing is created. The sharper the edge, the sooner aliasing is created. Similarly, if a horizontal or vertical edge moves between two temporal samples, temporal information is created. Fast motion creates temporal aliasing and, if the spatial edge is sharp, temporal aliasing is created even by slow motion.

In Chapter 5 we looked at the behavior of the two-dimensional DCT under these conditions. A horizontal edge with constant vertical position is handled well, but if the edge moves enough to create a vertical alias, DCT fails. The same happens if we try to extend DCT to image sequences; movement of a spatial edge in time creates temporal artifacts, and DCT is no longer effective.

The fundamental problem is, of course, that the temporal sampling rate of today's imaging systems is too slow. If an object moves significantly between two samples, we have no information about what happened between the two samples. In fact, we do not even know for sure that it is the same object! As discussed in Chapter 2, the human psychovisual system usually does an excellent job of "smoothing" the sequence of samples—based on its understanding of the world in which we live. However, the wagon wheel example demonstrates how this process is based on assumptions, and that these assumptions can be wrong. The reverse-motion effect is caused by making the wrong assumption about which spoke is which.

To determine correlation between samples at different times, and to permit temporal prediction, we need to build a signal-processing system that tries to emulate the processes of the human brain. We do not have anything like the same processing power as the brain, but we must do the best we can. Inevitably, any system we choose will make mistakes, probably more than our brains would.

The Motion Compensation Approach

Static Backgrounds

Many "moving" images or image sequences consist of a static background with one or more moving foreground objects. In this simplified case it is easy to see how we can gain some coding advantage. Figure 8-1 shows part of two temporally adjacent images from a sequence, and the division of each into blocks of pixels. Most of the image is unchanged from one time instant to the next, but a foreground object moves. Suppose we encode the first image using DCT and quantization, as with baseline JPEG. We transmit that image, and reconstruct it at the decoder. Now let's look at the second image, but keep the first image in store at both the encoder and decoder. We refer to this as the *reference image*.[1]

Figure 8-1
Moving foreground on static background.

We now treat the second image one block of pixels at a time, but before performing the DCT, we compare the block with the same block in the reference image. If the block is part of the static background, it will be identical to the corresponding block in the reference image. Instead of encoding this block, we can just tell the decoder to use the block from its copy of the reference image. All we need is one special code, and we can avoid sending any data from the background blocks. Where the pixel block in either image includes part of the moving foreground object, we will probably not find a match, so we can encode and transmit that block using DCT and quantization just as we did with the first image.

The benefit gained from this approach obviously depends on the picture content, but it is certainly possible that in an image with a

---

[1] In a real system, we have to be slightly more clever than this. For the image to be most useful it needs to be identical at encoder and decoder. A good design encodes the first picture and sends it to the decoder; at the same time, the transmitted data is decoded in the *encoder* to form a reference image identical to that derived by the decoder. This is similar to the technique described in Chapter 6 for predictive coding with quantization. Once there is a loss in the system, prediction must be based on reference data common to both encoder and decoder. To achieve this we emulate the decoder inside the encoder. The technique is described in Chapter 9.

static background half or perhaps three-quarters or more of the blocks will need no coding other than the "same as previous image" code. The data for transmission will be substantially reduced and, in this special case, we will have gained a substantial advantage from the temporal correlation of the images.

Motion Vectors

The previous section shows that with a static background we could expect to obtain a significant coding advantage, but this is clearly a very special case. Let's consider a similar case, but where the camera pans slightly to one side between the two images (see Figure 8-2). Now if we try the test of comparing a block to the corresponding block in the previous image we will not get any matches.

Figure 8-2
Moving foreground,
background shift.

However, we know that for most of the blocks the data exists in the previous image; it is just not in quite the same place. If we could measure the displacement we could still instruct the decoder to get the block of pixels from the previous image. We must send to the decoder the instruction to use data from the previous image, plus a *motion vector* to inform the decoder exactly where in the previous image to get the data. It would be sensible to make this a relative measure, so that the motion vector would be zero for a static background. This procedure is illustrated in Figure 8-3. The motion vector is a two-dimensional value, normally represented by a horizontal (x) component and a vertical (y) component. The process of obtaining the motion vector is known as *motion estimation*; using the motion vector to eliminate or reduce the effects of motion is known as *motion compensation*.

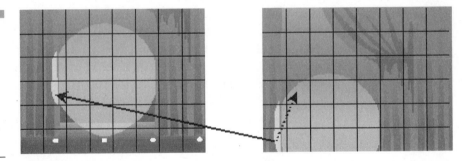

Figure 8-3
Generation of a
motion vector. Solid
arrow shows like
content found in
pervious image;
dotted arrow shows
actual magnitude of
motion vector.

Figure 8-4 shows part of a monitor display of an image with motion vectors superimposed. The camera is tracking the horse and rider, so background motion vectors are long and horizontal. On the rider, several vertical vectors show the rise from the saddle. Generally the figure shows that, in large areas of the picture, the motion vectors are well correlated, but where motion is most complex, the vectors appear almost random. This image was produced by a Snell & Wilcox MVA100 video analyzer for MPEG2 encoding and decoding; my thanks to Roderick Snell for permission to use it.

Figure 8-4
Motion vectors
superimposed on an
image. Photograph
courtesy of Snell &
Wilcox.

Block Matching

Static backgrounds and moving backgrounds provide a simple visualization of how motion vectors may be used to identify correlation between images in a sequence. However, it is evident that there is nothing special about the background—this technique can be used on any part of the image provided we can find the same object, or part of an object, in two images of the sequence. The simplest technique, *block matching*, is to treat each image block separately and search for an identical or similar block in the reference image.

The difficult bit is finding the matching block. What makes it difficult is the same reason we are looking at this problem in the first place—the lack of adequate temporal sampling. If the temporal sampling obeyed Nyquist, we would have a very easy task in tracking an object from one sample to the next. As we saw in Chapter 2, Nyquist sampling implies the simplest possible path that connects an array of samples. But we have already established that the temporal sampling of any common imaging system is much less than Nyquist. This fact leads to a simple conclusion: given the position of an object in one image of the sequence, we have no idea where it will be in the next!

Matching Criteria

Before we start to search for something, it is a good idea to establish the criteria by which we will decide that we have found it. In the previous sections I naively assumed that there would be a block identical to the one we were trying to match. In practice we do not want to look for just the perfect case; the object may have moved by some distance other than an integral number of pixels or there may be slight lighting variations. In fact, there are a large number of reasons why we may not get an exact match, and we must be able to make use of a close match. This means we must find a way of encoding the error.

Most motion estimation schemes look for a *minimum mean square error* (MMSE) between blocks, or for the slightly less complex *minimum absolute difference* (MAD) of two blocks.

Optimum Block Size

It is also important to consider the effect of the block size that we attempt to match. Previously I just assumed that we would search for a matching DCT block, but in practice this may not be the best choice.

In some ways it is obviously good to use a large block; in this way we can predict a large part of the image with a single motion vector, and the compression efficiency should be very high. However, a large block may contain a small moving object. The larger this object, the less likely we will find a predictor block that matches reasonably well; we may find no acceptable match or we may arrive at a distorted (wrong) motion vector. If the object is very small, it may not significantly disrupt the search, but the mismatch between blocks may be difficult to encode. In fact, the larger the block, the less likely that our matching criteria (MMSE or MAD) will find blocks whose differences are easy to encode.

If, on the other hand, we use a very small block, it is most unlikely that it will contain anomalies. If the MAD or MMSE is low, it is very likely that any difference between the blocks will be easy to encode. However, with small blocks we will have to encode many more motion vectors. It is also very likely that many matches will be found for a very small block. Most of these will not be related to real motion within the image; they will represent other areas where there happens to be a similar patch of image. This means that the "motion vectors" we calculate will likely be largely random rather than well correlated, thus the motion vectors themselves will not encode efficiently.

At the limit, if we were to take a block the size of a single pixel, any other pixel in the search range with the same intensity value would give a "motion vector" providing a perfect block match. Unfortunately, instead of encoding an array of well-correlated one-dimensional pixels, we would now have to encode a set of poorly correlated two-dimensional motion vectors.

The entropy and variance of the motion vectors can be used to evaluate the performance of block matching for various sizes of block. As might be expected, very large and very small block sizes lead to high-entropy and high-variance motion vectors. In fact, for most normal imagery, it can be shown that the 8×8 block used for the DCT is too small to be a good block match reference; 12×12 is also too small; 16×16 is the first easy-to-use size that gives good correlation of the resulting motion vectors and is, therefore, the appropriate choice in the absence of other factors.

The big question, however, is how to go about finding the matching blocks.

Full-Search Block Matching

Full-search block matching tests every possible block within a defined search range against the block it is desired to match. The technique is accurate and exhaustive—if there is a match within the search range, this technique will find it. It is, however, computationally demanding. Let's look at some numbers, assuming a 16x16 block size, known in JPEG and MPEG as a *macroblock*.

How do we decide on the search range? This depends on how fast we anticipate that objects will move in the image sequence, and how determined we are to track the fastest objects. There is no possibility of tracking an object that moves more than a picture width between successive images—if it is there in one image it will not be there in the next. It might be reasonable to think of tracking an object that moves half a picture width between successive images, but in television terms this corresponds to an object that appears, crosses the screen, and disappears in less than one-tenth of a second—probably faster than we need. If we aimed at tracking an object that traverses the screen in about half a second we would need to accommodate a displacement of about 50 pixels/image in standard-definition television. If we wish to predict for, say, three frames, that means a search range of about 150 pixels in each direction. For high-definition television we should probably double these numbers.

In real-world scenes, there is usually more or faster motion horizontally than vertically, and research has shown that for a given search area it is optimal for the width to be about twice the height.

How realistic is this? It suggests a search range of ±150 pixels $\times$ ±75 pixels, or a total search area of 300 pixels x 150 pixels. This does not sound too bad until we take into account the fact that for a motion vector resolution of one pixel we have to search for a matching block in every pixel position. For each quadrant of the search range this means $150 \times 75 = 11{,}250$ comparisons. In total, including the zero displacement case, we have 45,001 comparisons. Each comparison requires computing the sum of the differences (or the squares of the differences) of 256 individual pixels. Taking the simpler case, this requires 513 calculations per block, or about 23 million calculations per search—and that is for just one block in the image we are trying to code. In a standard-definition television picture, this means over 30 billion calculations per frame, almost 1 trillion calculations per second. Some compression schemes permit half-pixel resolution for the motion vectors, requiring the calculation of interpolated pixel values as well as the additional comparisons at the new positions. In the example quoted, this probably brings the calculation requirement to over 1 trillion per second, just for standard-definition 720x480 video at 30 frames/s. Block matching does not come cheap!

Although the approach can be viewed as rather simplistic, full-search block matching will usually find the best match within the search area. However, the best match, as determined by MAD or MMSE, may not necessarily represent the "real" match—that representing the true motion in the scene. Good matches will minimize the need for transmission of *residual errors* (see below), but if the matches do not represent true motion, the motion vectors of adjacent macroblocks may not correlate well, and these vectors will be costly to encode.

There are some techniques that can increase the probability that a match will represent true motion. If the are successful, this reduces the entropy of the motion vectors, increasing overall compression efficiency.

Hierarchical Block Matching

Hierarchical block matching seeks to take advantage of the use of larger blocks by filtering to eliminate fine-detail information. The idea is to use the best-match position of a large fuzzy block to arrive at a first approximation of motion. The approximation can then be successively refined by searching smaller areas using smaller block sizes, each block size being appropriately filtered.

With this technique, the likelihood of a finding the "wrong" motion vector is decreased substantially. The general trend of motion is established by the large filtered blocks, and the accurate measurement is accomplished by using small blocks. It has been shown that hierarchical block matching on real images yields motion vectors with significantly lower variance and entropy than those uncovered in a search using any single block size.

Hierarchical block matching requires many filtering operations that in themselves demand a great deal of computational power, but with appropriate design the number of block comparisons can be considerably lower than with full-search matching.

Motion estimation is a very important part of asymmetrical compression systems—systems where there are few encoders and many decoders—and generally it is well worth additional cost at the encoder to achieve even a small improvement in compression efficiency. As more computational power becomes available to encoder designers, motion estimation systems will inevitably become more complex and sophisticated. Some possible directions are discussed below. But, before we examine these alternatives, we must look at what happens when the best match we can find is less than perfect (as will frequently be the case).

Residuals

Earlier, we considered the trivial case of encoding a sequence where the background was static. Here, either we found a match where two pixel blocks were identical, or we did not. Using block matching, we recognized that in the real world we would have to search for the best match we could find, and that this match may not be perfect.

There are many possible reasons for an imperfect match; the object being tracked may have moved closer or farther away, or rotated, or there may have been a lighting change. (And despite all our efforts, the match may be coincidental, merely representing two similar blocks in the image.)

An imperfect match does not necessarily mean that we cannot benefit from motion compensation. We can determine the difference between the block of pixels we want to transmit and the best match we can find. The differences are known as *residuals*. The residuals may be encoded and transmitted along with the motion vector. At the decoder the residuals are decoded, added to the pixel block referenced by the motion vector, and the block is reconstructed.

Of course, if the match is not very good, it may take more bits to transmit the motion vector plus the residuals than to code and transmit the pixel block itself; a good coder tests both and uses the most efficient mechanism.

Today most encoders use MAD or MMSE to test for the best block match. Although these are simple and cheap tests for the match, they do not give any meaningful measure of how much data may be needed to encode the residuals. Future encoders may use more computational power to choose the motion vector based upon the lowest requirement for residual encoding.

Other Motion Estimators

Restricted Search Systems

The techniques of motion estimation described above are brute-force ones and require a great deal of computational effort. Obviously, there are other approaches. For example, we could use the motion vector from the previous block as a starting point for a search. Alternatively, we could test for a match at a few points on a coarse search grid, and use the best match as a starting point for a search on a fine grid, but over only a small area.

Some of the alternative techniques will find the best match as certainly as the full-search approaches and are likely, but not certain, to do so with a lower number of calculations. Others will always use fewer calculations, but are not guaranteed to find the best match.

These characteristics give a clue to possible applications. High-grade software encoders must find the best match, but run faster if, at least most of the time, they look in the right place first. Other forms of software encoder, such as might be used for video conferencing, operate on a real-time signal just like a hardware encoder. They have to achieve results in a short time or give up (skip a frame), and the search algorithm is just one of the many compromises in the design of a smart encoder.

This argument shows why such techniques are not often employed in high-grade hardware encoders. These encoders are not allowed to skip frames or perform any tricks. Therefore, they must be capable of carrying out the maximum calculation load every frame. Given this condition, there is little to be gained by systems that reduce the computational load for *some* frames—the hardware has to be there, so it might as well be applied in the most straightforward fashion for every frame.

In fact, the trend for hardware encoders is in the opposite direction. In asymmetric systems, the coder cost is paid only once; this is in contrast to what happens in large numbers of transmission systems and/or decoders. Computational power gets cheaper every year, and increased computational power can lead to better coding efficiency. For any given system, better efficiency means fewer bits (and a longer movie on the disk) or higher quality. It can be expected that increasingly sophisticated techniques will be applied to encoders for such systems. One approach is described in the next section.

Phase Correlation Motion Estimation

Phase correlation is a means of determining very accurately the components of motion in an image sequence. Phase correlation soon became the process of choice for standards conversion, where the requirements for accurate motion estimation are even more stringent (see below). The use of phase correlation for motion estimation in an MPEG encoder was first demonstrated by Snell & Wilcox in September 1997.

The need for standards conversion between different frame rates is another problem resulting from the sub-Nyquist temporal sampling of moving images. Standards conversion requires that we start with one set of temporal samples of a moving scene and generate a different set, corresponding to the new frame rate. If the original samples were close enough together (Nyquist sampling) this would be a trivial interpolation exercise. Unfortunately, for all the reasons discussed in this book, this simplistic approach does not work. A standards converter can interpolate static and slow-moving parts of the image, but to be able to place fast-moving objects correctly in the new frames the objects must be tracked, just as we have been discussing in this chapter.

In fact, the job of the motion estimator in a standards converter is far more difficult than that in a compression encoder. In the compression device we may detect a block match that is coincidental, not corresponding to the true motion of any object in the scene. If this happens, we use a motion vector that is not correlated with any of the true motion vectors, thus it is less efficient to encode. Other than that, we do not care. Even though the motion was spurious, the predictive coding of the image block is almost as efficient as if we had found true motion. In a standards converter, a mistake like this means that a frame is produced with a false object, or an object in a place that it should not be. Such errors are, of course, very obvious.

Phase correlation was developed to measure motion accurately in an image sequence, thereby avoiding the type of error described above. It is a complex technique, and I attempt only the briefest of explanations here. John Watkinson (Watkinson, 1994) gives a more detailed description, together with a nonmathematical explanation of the Fourier transform.

We looked briefly at the Fourier transform in Chapter 5. It is a fundamental process from which the DCT was derived. Basically, the Fourier transform moves between the spatial and frequency domains. When we perform a Fourier transform on a block of spatial data (pixel values) we get a block of frequency data, each frequency characterized by an amplitude and a phase.

In the phase correlation motion estimator, blocks from adjacent fields are Fourier-transformed and the transforms are subtracted. Assuming similar frequency content, the main components of the result are frequencies where there were a significant phase shift between the fields. All these frequencies are normalized to the same amplitude and passed to an inverse Fourier transform. This generates a new value associated with each pixel, and these values can be interpreted as a spatial surface (called a *correlation surface*). This surface displays amplitude peaks corresponding to the movement of each object. The position of the peak gives an accurate measure of the direction and speed of motion.

Unfortunately, having found a technique for accurately measuring the motion, we have done so by working through the frequency domain, and we have lost all precision in the spatial domain. Phase correlation tells us very accurately the amplitude and direction of each motion component in the image, but it does not tell us where in the image the motion occurs.

We can find the area of motion by taking the two blocks of pixel data and offsetting them by the amplitude and direction corresponding to the phase correlation peak, and then subtracting. The area of image that exhibits that particular motion is nulled in the result of the subtraction.

Not every peak in the correlation surface necessarily corresponds to real motion of part of the image; spurious peaks can also be generated. The technique of offsetting and subtracting allows the area of each true motion to be identified, and eliminates any spurious results.

Ideally, phase correlation is performed over the whole picture area, but the amount of computation required makes this impractical. Instead, the process is applied to a number of overlapping windows. The size of the window, like the size of the search area for block matching, determines the maximum motion speed that can be handled. Again, this is an area where the future will bring substantial improvement through the availability of more processing power.

MPEG-1

Introduction

MPEG stands for Moving Pictures Experts Group and, like JPEG, it is a committee formed under the Joint Technical Committee of ISO and IEC discussed in Chapter 7. The committee was formed in 1988 under the leadership of Dr. Leonardo Chiariglione of Italy. Things started out slowly; attendance at the early meetings was about 15 delegates, and when two consecutive 1988 meetings were held in Torino and London, only two delegates attended both sessions. This soon changed, and by early 1990 attendance had grown to over 100 delegates at every session.

The first task of the committee was to derive a standard for encoding motion video at rates appropriate for transport over T1 data circuits and for replay from CD-ROM—about 1.5 Mbits/s. A measure of the aggressiveness of this objective may be seen by looking at the numbers for an audio CD. A regular audio CD, carrying two-channel audio, at 16-bit resolution with a sampling rate of 44.1 kHz, has a data transfer rate of more than 1.4 Mbits/s. The MPEG-1 goal was to record the same audio, but sampled at 48 kHz, *together with video,* at the same data rate!

TABLE 9-1

MPEG-1
constraints

Parameter	Constraint
Horizontal picture size	≤768 pixels
Vertical picture size	≤576 lines
Number of macroblocks	≤396
Number of macroblocks/s	≤396 × 25 = 9900
Picture rate	≤30 pictures/s
VBV buffer size	≥2,621,440 bits
Bit rate	≤1,856,000 bits/s

To focus the design of the system around a practical objective, certain parameter constraints were adopted, as shown in Table 9-1. These parameter values represent boundaries; a bit stream with any parameters outside these boundaries is not MPEG-1, and an MPEG-1 decoder is not required to decode it.

The most significant numbers here are the number of macroblocks and the number of macroblocks per second. It might be thought from a glance at Table 9-1 that MPEG-1 could encode standard-definition television, but this is not the case. If we take the pixel limits,

$$\frac{768}{16} \times \frac{576}{16} = 1728 \text{ macroblocks/picture}$$

the real limit is set by the limit of 396 macroblocks per picture, and by the limit of 396×25 macroblocks per second. These correspond to an image size of 352×288 for 50-Hz systems and 352×240 for 60-Hz systems. They both give, 9,900 (396×25) macroblocks per second. This format, in its two variants, is known as *common intermediate format* (CIF), and is the image size used by many non-broadcast-quality compression systems. It represents half the horizontal resolution and half the number of lines of standard-definition television. However, it is quite a close approximation to the delivered picture quality of interlaced NTSC on a typical domestic receiver, except for the lower temporal resolution.[1]

MPEG-1 (and, as we shall see in the next chapter, MPEG-2) differ from JPEG in two major aspects. The first big difference is that both permit temporal compression as well as spatial compression. The second difference follows from this—temporal compression requires three-dimensional analysis of the image sequence, and all known methods of this type of analysis are computationally much more demanding than two-dimensional analysis. Motion estimation is how we approach three-dimensional analysis today and, as we saw in Chapter 8, practical implementations are limited by available computational resources. MPEG-1 and MPEG-2 are both designed as asymmetric systems; the complexity of the encoder is very much higher than that of the decoder. They are, therefore, best suited to applications where a small number of encoders are used to create bit streams that will be used by a much larger number of decoders. Broadcasting in any form and large-distribution CD-ROMs are obviously appropriate applications.

As with JPEG, Joan Mitchell and William Pennebaker (with other experts) have produced a detailed text on the MPEG standards that is a must for implementers (Mitchell, 1996). This chapter provides only a brief summary of the standards.

What MPEG Defines

The previous section gives a clue about the model adopted by MPEG for standardization. MPEG *describes* various tools that may be used to

---

[1] The arguments about the advantages and disadvantages of interlace will never be resolved to everyone's satisfaction, but all interlaced television cameras do provide temporal resolution equal to the field rate.

perform compression and gives some *examples* of how these might be implemented. MPEG *defines* the syntax of a compliant bit stream and the ways in which a decoder must *interpret* valid bit streams, those that conform to the defined syntax.

The really important thing here is the omission. MPEG does *not* define the encoder. A valid encoder is any device, implemented in hardware and/or software, that produces a syntactically correct bit stream, resulting in the desired output if the bit stream is fed to a compliant decoder. This is a powerful approach, because no restrictions apply to techniques or technologies used in the encoder. Any process, any algorithm, may be used in the encoder, provided the output can eventually be expressed by some combination of the permitted syntactical elements. Given this, no matter what advances are made in encoder technology, any compliant decoder will produce the correct output.

Note that this approach to standardization allows for a wide variety of implementations. I commented in the previous section that MPEG-1 and MPEG-2 were designed as asymmetric systems. This is true, but the way the standard is written allows us to deviate from this if we wish. It is quite possible to build an MPEG encoder without any capability for temporal compression. The complexity of the encoder is then very similar to that of the decoder (as in JPEG). Provided the output of the encoder includes *only* syntactically valid elements, there is no requirement to use all the tools. An MPEG encoder that performs spatial compression only is perfectly valid if it stays within the rules of the syntax, and its output will be decoded by any MPEG compliant decoder.

At the decoder, the rules are different. An MPEG-1-compliant decoder is required to decode all valid MPEG-1 bit streams. Again, within a defined system, it is perfectly possible to build encoders and decoders that (for example) do not support temporal compression. As discussed above, such an encoder is MPEG-1-compliant provided its output bit stream includes only valid elements, but a decoder built this way is *not* MPEG-1-compliant; to be described as compliant it must be capable of decoding *any* valid MPEG-1 bit stream.

Hierarchy and Terminology

The top-level definition in MPEG-1 is the *sequence* of pictures. A sequence is of arbitrary length and can represent a video clip, a complete program item, or a concatenation of programs.

Within the sequence, the next lower definition is the *group of pictures* (GOP). The GOP is a very important concept in MPEG and is discussed at length below. A typical MPEG bit stream consists of a repeating GOP structure. In the simplest form of encoding, without temporal compression, the GOP can be a single picture. However, in typical MPEG applications, the GOP includes pictures coded in three different ways and arranged in a repetitive structure, most commonly between 10 and 30 pictures long. We will look at some typical GOP structures after examining the different types of frames that may be used.[2]

This brings us to the *picture*, or *frame*. This is fairly obvious, but there is one important point to make. MPEG-1 includes no concept of interlace; every picture or frame contains all the lines that make up a complete image in the sequence. This is not the same as saying that MPEG-1 cannot encode interlace. The two fields of an interlaced frame can be combined as one frame, but MPEG-1 makes no concession to this practice. There are no special tools for interlace, and no means of avoiding the coding inefficiencies that may result from interlace (in an interlaced system the frame consists of lines that were captured at different points in time).

Continuing down the hierarchy, we reach the slice and the *macroblock*. A macroblock contains all the information required for an area of the picture representing 16×16 luminance pixels. Macroblocks are numbered in scan order (top left to bottom right). In MPEG-1, a slice is any number of sequential macroblocks (Figure 9-1). The chief significance of the slice is that it is encoded without any reference to any other slice; if data is lost or corrupted, decoding and recovery can usually commence at the beginning of the next slice.

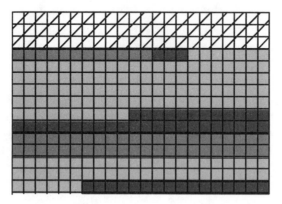

Figure 9-1
Slices in MPEG-1.

---

[2] MPEG-1 defines a fourth picture type, the D-frame. This is a low-resolution (thumbnail) type of frame that cannot be used in conjunction with other frame types. It is rarely used and is not discussed further in this book.

Frame Types in MPEG

Intraframes (I-frames)

The prefix "intra" means *inside* or *within*, and an *intraframe* or *I-frame* is a frame that is encoded using only information from within that frame. In other words, it is a frame that is encoded spatially with no information from any other frame—no temporal compression. Coding of an I-frame is similar, but not identical, to coding of an image in JPEG or a single frame in motion JPEG.

Intra is such an important concept in MPEG that it is used as a word rather than just as a prefix. The intraframe is intracoded, as opposed to frames that use information from other frames that are described as *inter-* or *non-intra*.

Non-intra Frames (P-frames and B-frames)

Non-intra frames use information from outside the current frame, from frames that have already been encoded. In non-intra frames, motion-compensated information, (as described in the previous chapter) is used for a macroblock where this results in less data than directly (intra) coding the macroblock.

There are two types of non-intra frames, *predicted frames (P-frames)* and *bidirectional frames (B-frames)*. These are described with reference to the series of images shown in Figure 9-2. The first image (X) shows a background, a tree, and a car. The third image (Z) shows the same scene somewhat later. The camera has panned to the right, causing all static objects to move left within the frame, but the car has moved to the right, and now obscures part of the tree. The second image (Y) shows a point in time somewhere between X and Z.

Image X is encoded as an I-frame; as we noted, this means that it is totally spatially encoded without reference to any data outside that frame. When the frame has been processed, the encoded data is sent to the decoder where it is decoded and the reconstructed image is stored in memory. At the same time, the same encoded data is decoded *at the encoder* to provide a reconstructed version of X identical with that stored at the decoder. We'll call the reconstructed version X*. This sounds strange, but we must remember that this is a lossy compression scheme, so that the frame reconstructed at the decoder may not be (and likely will not) be identical to the original. We are going to use

this frame as a reference, and the only version available at the decoder is the reconstructed frame. It is, therefore, essential that we create an identical frame at the encoder to be the reference. This process is shown in Figure 9-3.[3]

---

[3] Using the original frame as a reference is, however, more likely to result in true motion being found. Test model 5 (see below) uses the original to determine motion vectors to 1 pixel accuracy, then the reconstructed frame (which should give more accurate residuals) to refine the motion vector to half-pixel accuracy.

Figure 9-3

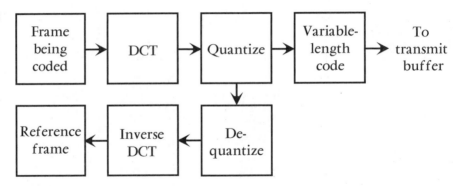

Reconstructing a
reference frame that
will be the same as at
the decoder.

Image Z will be encoded as a P-frame. For each macroblock in Z the encoder will search for a matching macroblock in X*. The objective is to find a motion vector that links the macroblock to an identical or very similar macroblock in X*. As described in the previous chapter, residuals must be considered if the macroblocks are not identical.

Assuming that the motion detector is good and has an adequate search range, we should be able to encode a large proportion of image Z with respect to X or X*. Notable exceptions are the area behind the car in X and the strip of image at the right-hand side of Z that was revealed by the camera pan—this information does not exist in picture X.

This gives a clue about how we might improve the efficiency of temporal encoding. In the image sequence shown, X is the earliest in time, followed by Y, followed by Z. We have described the coding of Z, but as yet we have not discussed how Y is encoded. Suppose we store picture Y away until after we have encoded Z. Now we can use both X* and Z* as reference frames while encoding Y. In this idealized case we should be able to find a motion vector for almost every macroblock in Y. This type of encoding is known as *bidirectional encoding*, and frame Y is said to have been encoded as a B-frame.

The coding of different frame types is discussed in more detail below, but having established the principles of unidirectional and bidirectional encoding, we can now examine the structure of the MPEG group of pictures.

MPEG Group of Pictures

A typical MPEG group of pictures is shown in Figure 9-4. The pictures are now categorized in a slightly different way. I-frames and P-frames are called *anchor frames*, because they are used as references in the coding of other frames using motion compensation. B-frames, however, are not anchor frames, because they are never used as a reference.

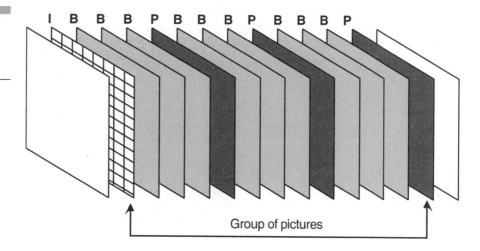

Figure 9-4
A typical group of pictures in display order.

I B B B P B B B P B B B P

Group of pictures

The GOP shown starts with an I-frame. Clearly, it is essential to *code* an I-frame first to start the sequence; if no previous information has been received there is no possible reference for motion estimation. It is possible to have a number of B-frames precede the I-frame because these are encoded and transmitted after the I-frame (see below). The first P-frame is coded using the previous I-frame as a reference for temporal encoding. Each subsequent P-frame uses the previous P-frame as its reference. (This shows an important point; errors in P-frames can propagate, because the P-frame becomes the reference for other frames.) B-frames are coded using the previous anchor (I or P) frame as a reference for forward prediction, and the following I- or P-frame for *backward prediction*. B-frames are never used as a reference for prediction.

Figure 9-4 shows a *closed GOP*, meaning that all predictions take place within the block. Many MPEG practitioners prefer this approach, but there are examples of *open GOP* structures such as I-B-I-B-I... . Such structures can be quite efficient, but there is no point at which the bit stream can be separated; every frame boundary has predictions that cross it. The GOP in Figure 9-4 may also be described as *regular*, in that there is a fixed pattern of P- and B-frames between I-frames. Regular GOPs may be characterized by two parameters, M and N. M represents the distance between I-frames, and N is the distance between P-frames (or closest anchor frames). It is possible to construct *irregular GOPs*, but these are not commonly used.

Note that B-frames can be decoded only if both the preceding and following anchor frames have been sent to the decoder. Figure 9-4 shows the GOP in *display order*, but to enable decoding, the frames are actually transmitted in a different order, as shown in Figure 9-5. It is important to note that this reordering, essential when B-frames are used, adds substantially to the delay of the system.

Figure 9-5
The same group of pictures as shown in Figure 9-4, but pictured in coding (and transmission) order. The numbers below the frames represent display order.

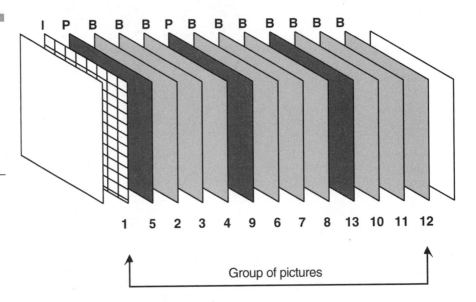

Coding of Macroblocks

It can be seen from the preceding discussion that the coding decisions for a macroblock depend upon the type of frame. However, before considering this, we must look at the structure of a macroblock and the various possible ways of coding it.

MPEG-1 uses Y, C_B, C_R coding and a 4:2:0 structure of color information.[4] This means that the luminance is coded in every pixel, but the color difference information is filtered to half the luminance resolution, both horizontally and vertically. Thus an image area represented by a block of luminance pixels 16×16 requires only 8×8 for C_B, and 8×8 for C_R. Because we will use 8×8 blocks for all DCT coding, the macroblock consists of four blocks of luminance samples and one block each of C_B and C_R samples (see Figure 9-6).

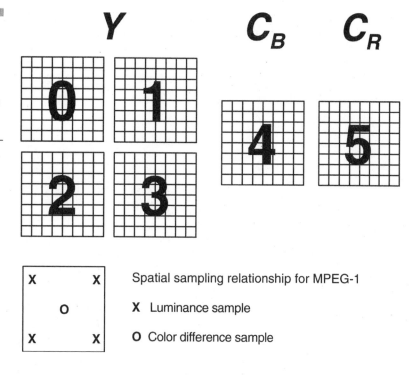

Figure 9-6
The MPEG-1 macroblock, the coding order, and the spatial relationship between luminance and color difference samples.

Spatial sampling relationship for MPEG-1

X Luminance sample

O Color difference sample

Intracoding of Macroblocks

When a macroblock is intracoded, the procedure for each block of samples is almost exactly the same as described for JPEG. Here we examine the significant differences. Unlike JPEG, MPEG defines default quantization tables; one is used for intracoding, the other is used to code any residuals when predicting by motion estimation. Either or both quantization tables may be changed by specifying new tables in the sequence header; otherwise the decoder assumes the

---

[4] The nomenclature is bizarre. Originally this type of resolution numbering was adopted to describe the digital representations of 525- and 625-line television. Early work had used multiples of color subcarrier frequency for sampling. Eventually the rate of 13.5 MHz was chosen for both standards, and because this rate was fairly close to four times the NTSC subcarrier frequency, it acquired the descriptor "4" to represent the luminance sampling frequency. The descriptor 4:2:2 means that Y is sampled at 13.5 MHz, and that C_B and C_R are each sampled at half this frequency—hence 4:2:2. The MPEG delegates either did not know this, or chose to ignore it. 4:2:0 does not mean zero sampling frequency for the C_R signal! In MPEG-ese, 4:2:2 means that both color difference signals have half the horizontal resolution of luminance; 4:2:0 means that the same two signals have half the horizontal *and* half the vertical resolution of luminance!

defaults. The default table for intracoding is shown below; for non-intracoding of residuals the table is flat—all values are 16.[5]

8	16	19	22	26	27	29	34
16	16	22	24	27	29	34	37
19	22	26	27	29	34	34	38
22	22	26	27	29	34	37	40
22	26	27	29	32	35	40	48
26	27	29	32	35	40	48	58
26	27	29	34	38	46	56	69
27	29	35	38	46	56	69	83

MPEG scales the quantization by means of a *quantization scale factor*, sometimes known as *MQuant*. This value is set explicitly for every slice, but may be changed at the macroblock level if required. Each macroblock is assumed to use the same value as the preceding one, unless a new value is set in the macroblock header. MQuant values range from 1 to 31. The AC coefficients are quantized by multiplying by 8, then dividing by the appropriate value from the quantization table, and by MQuant. An MQuant value of 1 decreases the severity of quantization by a factor of 8 (higher quality); a value of 31 increases the severity by a factor of 4 (approximately), yielding a higher compression ratio, but lower picture quality.

As in JPEG, the DC coefficients are extracted and coded predictively. The AC coefficients are quantized as described above, then scanned and variable-length-encoded almost exactly as described for JPEG.

Non-intracoding of Macroblocks

In a P- or B-frame, the first step in coding a macroblock is to intracode it, as if we were coding an I-frame. If we fail to find a reasonable

---

[5] It should be noted that Test Model 5 discussed below uses the default intra quantization matrix, but does not use the default flat matrix for residuals. This may suggest that experience has shown significant benefit in deviating from the system default in this case.

match in the motion estimator, or if the best match requires transmission of substantial residuals, we revert to intracoding. The motion estimator is used to find the nearest match to the macroblock, either in the preceding anchor frame (for P-frames), or in both the preceding and following anchor frames (for B-frames). In MPEG, only the luminance samples are used in motion estimation, but the resulting motion vector is used in coding both luminance and color difference blocks.

When the best match has been found, each block within the macroblock is treated separately. Each block being encoded will be subtracted (pixel by pixel) from the corresponding block in the matching macroblock, leaving an array of error values. This block of error values (residuals) will be encoded by DCT and quantization as in intracoding, except that the flat interquantization matrix is used. In intercoding, the residual DC coefficient is not treated separately, but quantized and scanned along with the AC coefficients. (In this case the DC coefficient results from a prediction. In fact, it is the mean absolute difference that was most likely used to select the best match in the motion estimator. The chances are very high that this coefficient will quantize to zero, so that there is no benefit in special treatment.)

This process is applied to the six blocks that make up the macroblock (four luminance plus two color difference). If the motion estimation is good, many blocks will likely produce small residuals that quantize to zero throughout; these blocks do not need to be encoded at all and can be skipped. When all blocks have been either encoded or skipped, a code is chosen to represent the pattern of blocks that have actually been encoded. If no blocks need encoding, the entire macroblock may be skipped.

We also need to code the motion vectors. For the first macroblock in a slice, the motion vectors must be sent in full. For the rest of the slice, the vectors are predictively coded; in other words, the vectors are assumed to be the same as for the preceding macroblock and differential values are sent for correction if required. As discussed in Chapter 8, if the motion estimator has found true motion, this predictor works well. If the motion estimator has merely found a random macroblock that happens to have similar pixel values, the prediction will be poor and many bits will be used for transmission. Both motion vectors and differentials are variable-length-encoded.

P-FRAMES

If we are encoding a P-frame, we can now make a choice. If we have determined that the entire macroblock can be skipped, the decision is

easy: we just write a "skip" code into the macroblock header and proceed to the next macroblock. Otherwise, we can total the number of
bits that will be required to transmit the motion vectors, the code
showing which blocks of residuals have been encoded, and the coded
residuals, and compare this total with the number of bits required to
transmit the macroblock intracoded. If the motion-compensated coding uses fewer bits, we transmit that; otherwise we mark the macroblock as intra coded and transmit that data. (If we do this, the next
macroblock assumes a value of zero for the preceding motion vectors
when predictively encoding its motion vectors.)

B-FRAMES

When encoding a macroblock in a B-frame, we have a few more calculations to perform. As with a P-frame, we calculate the number of
bits for intracoding, and the number for intercoding with forward
prediction. Because this is a bidirectionally encoded frame, we do two
more things: the motion estimator finds a backward prediction, and
the encoder must go through exactly the same process as described
above to determine a total number of bits for intercoding with backward prediction.

Finally, the encoder may assess a fourth method of encoding that is
allowed for bidirectionally encoded macroblocks: *interpolated prediction*. For regular motion prediction, we have used the macroblock at
the end of the motion vector as a predictor. We subtract this from the
macroblock being coded and then deal with the residuals. In interpolated prediction we take the mean (on a pixel-by-pixel basis) of the
macroblock at the end of the forward motion vector and the macroblock at the end of the backward motion vector. We use this mean
as the predictor and subtract this from the macroblock being encoded
to form the residuals. So, if C is the current macroblock, F is at the
end of the forward motion vector, B is at the end of the backward
vector, and R is the macroblock of residuals:

$$R_{i,j} = C_{i,j} - \left(\frac{F_{i,j} + B_{i,j}}{2} \right)$$

where i and j range from 0 to 15 to cover all 256 samples.

Note that this is an unweighted average, irrespective of the position
of the B-frame with respect to the surrounding anchor frames. The
mechanism is, apparently, useful for encoding noisy picture
sequences.[6]

Any one of the intercodings may allow the macroblock to be skipped, although in this case we must specify which of the prediction modes provided that condition. Otherwise, the bit totals for the three prediction modes are compared. If the best of these needs fewer bits than intracoding, then that mode is used; otherwise the intracoding is transmitted.

Rate Control

All of the preceding techniques are described by MPEG, and the syntax of the bit stream is specified in exhaustive detail. As mentioned at the beginning of the chapter, MPEG specifies the decoder but not the encoder—this is left to the ingenuity of the implementer. Perhaps the blackest of the black arts associated with MPEG is rate control.

Consider the GOP structure. Even if there were no issue in distributing bits throughout a picture, the different coding methods create a problem in rate control. We must have intracoding, otherwise there is no place to start when creating video in the decoder. Obviously intercoding must be more efficient, or we would not go to the enormous difficulty and expense of implementing it, so we assume that a P-frame typically uses fewer bits than an I-frame, and a B-frame should use fewer still. This means that over a transmission system that uses a constant or nearly constant bit rate, I-frames will take longer to transmit than P-frames, which will take longer than B-frames. However, at the encoder, the unencoded frames are arriving regularly every one-thirtieth of a second, and the decoder must send a frame to the display every one-thirtieth of a second. In turn, this means that the delay from encoder input to decoder output must be constant. This can only be achieved with a buffer at each end.

Actually, the situation is far more complex than this. Even if we were to transmit all I-frames, we would want some form of rate control. Individual pictures vary greatly in their information content or entropy. A frame captured during a rapid pan of a correctly adjusted television camera contains little but blur. Even though there may be recognizable images, there is no detail or high spatial frequency content. For every block, a DC level plus a simple horizontal and/or vertical slope should

---

[6] However, one must wonder if the original intention was lost. A weighted interpolation, based on the temporal distance of the two anchor frames, would make a useful predictor for use during a dissolve or fade—something MPEG should have, but does not!

give a very close approximation to the image. In DCT terms, we should be able to represent each block by a DC coefficient and at most two or three AC coefficients. Such a picture needs very few bits to encode, whereas a static shot of a very detail-rich scene may require many more bits for an acceptable result.

Even with I-frame-only encoding, constant bit rate is likely to be satisfactory only with very high bit rates, such as may be used in a studio environment. Constant bit rate becomes variable-quality when we have images of varying complexity. Variable quality is very disturbing to a viewer, usually more disturbing than constant but mediocre quality. What we want is essentially constant quality for images of variable complexity. This in itself requires variable bit rate.

Within each picture a similar argument may be applied. Usually, some parts of a picture are more complex than others, and the bits must be used where they will do most good. Flat sky is very easy to encode, even easier than the blurred picture. If we allocate half the bits to the top half of an image and this is mostly sky, we will certainly be short of bits when we come to a complex foreground. If we start at the top of an image and allocate all the bits necessary to perform a high-quality encode of detailed background trees, again we will be short of bits for the foreground.

So, to do the job properly, we need to allocate bits in a medium- to long-term manner according to the complexity of the frames or sequences. We need to weight this allocation to distribute the bit allocation correctly among I-, P-, and B-frames, and within each frame we must distribute the bit allocation according to the distribution of complexity within that frame.

We would be justified in expecting an array of formidable tools to help us control bit allocation. In fact, we have one tool—the *quantization scale factor*, or MQuant. Everything else we do minimizes the number of bits for each macroblock for a given MQuant, but nothing else gives us any type of control over where the bits are used. Figure 9-7 shows the rate control mechanism of a basic MPEG encoder.

The encoder produces a variable-rate stream of bits that go into a buffer. The buffer is emptied by a constant transmission rate. If the encoder produces bits too slowly, the buffer will empty, and there will be nothing to transmit. If the rate from the encoder is too high, the buffer will overflow and data will be irretrievably lost. To try to stay within these two bounds we feed a measure of buffer fullness to the magic box called the rate controller, which responds by varying the quantization scale factor, affecting quantization of the DCT coefficients of blocks that have not yet been coded!

Figure 9-7
Rate control in
MPEG.

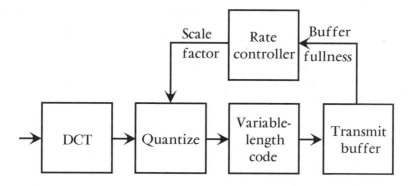

MPEG does provide an approach to the problem, called *Test Model 5 rate control*. TM5 starts by assuming a constant number of bits per GOP. It then makes a tentative distribution of bits based on the "normal" relative requirements of the different frame types. (The initial guess in TM5 is a ratio of 140:52:36 for I-, P-, and B-frames.) After each frame has been encoded, TM5 calculates the bits remaining within the GOP and recomputes the target bits/frame based on the new situation. It can also modify the ratios given above, on the basis of the actual performance of the encoder with the current input.

Within each frame, TM5 estimates, from known constants and recent history, the scale factor likely to generate the target number of bits. Then, after coding each macroblock, it reestimates the nominal scale factor from the number of bits used and the number remaining to meet the target.

Before the coding of each macroblock, the pixel values are examined to generate a measure of spatial activity within that macroblock. This measure is used to modify the nominal scale upward or downward by a factor of up to 2:1. A "quiet" macroblock will have the scale factor raised to conserve bits, a "busy" macroblock will be allowed more than its nominal share of bits to attempt to maintain good quality.

There are more sophisticated techniques—in fact, TM5 is regarded by today's designers as obsolete and simplistic. Some encoders calculate the spatial activity of each macroblock in the picture before starting to encode it. This allows a more accurate distribution of the target bits within each picture. Given this mechanism, some extra delay allows the spatial activity of each picture to be compared with its neighbors, and the target bits/picture can be varied accordingly. Even more sophisticated techniques may look for scene changes, or radical changes within a scene and predict the failure of motion compensation. The system then reduces the factors separating I-, P-, and B-frames. Note that all these techniques increase the delay of the system, and this may not be acceptable in some situations.

Even with these techniques, some facilities prefer to use a human "compressionist" for critical jobs like transferring movies to CD-ROM. The compressionist can use a two- or three-pass technique to plot the complexity of the scenes and tune the encoder parameters to maximize performance. A simple scene might be bit-starved toward the end to take the buffer to the absolute minimum pending a switch to a complex scene. In extreme cases, to cope with very complex motion, the compressionist may spatially filter the images to reduce the spatial complexity. MPEG is so complex, and our tools for objective quality measurement are so poor, that human intervention may provide a noticeable improvement in the subjective quality that can be obtained with a fixed number of bits.

We have talked about buffer fullness in the encoder. If this were the only issue, the encoder designer could avoid many problems by increasing the size of this buffer, provided the additional delay could be tolerated. However, we are using the buffer to convert variable rate to fixed rate and this process must be mirrored at the decoder. The sequence header of an MPEG stream includes a value for decoder buffer size (or, more accurately, for a particular mathematical model of the decoder buffer) known as *video buffer verifier* (VBV). The maximum size is specified by the MPEG standard (320 kbytes in MPEG-1). The encoder rate controller must track the fullness of this buffer, and ensure that it is never allowed to underflow or overflow. (The encoder designer must ensure that the mathematical model never underflows or overflows; the decoder designer has to ensure that the physical buffer matches the performance of the model.)

Figure 9-8 shows a simple example of VBV modeling. Bits arrive at a constant rate (hence the constant slope of the ramps), but are removed in variable-size blocks, corresponding to complete frames (of I, P, or B type).

Figure 9-8
VBV modeling in
MPEG.

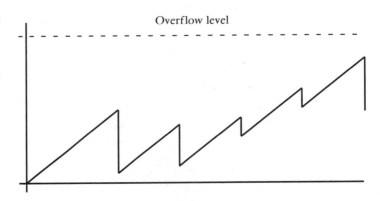

Overflow level

The MPEG Encoder and Decoder

Figures 9-9 and 9-10 show simplified diagrams of a basic MPEG encoder and decoder, respectively. In the encoder, the motion predictor compares the frame being coded to a reference frame (or two reference frames for bidirectional encoding). When a match is found, a motion vector is generated, specifying the location in the reference frame of the macroblock to be used for prediction. Either the macroblock, or the block of residuals formed by subtracting the predicting macroblock, is passed to the spatial encoder. This spatial encoder is very similar to the JPEG encoder discussed in Chapter 7, but with the addition of a rate-control loop. Motion vectors are coded predictively, in a manner similar to that used for the DC coefficients.

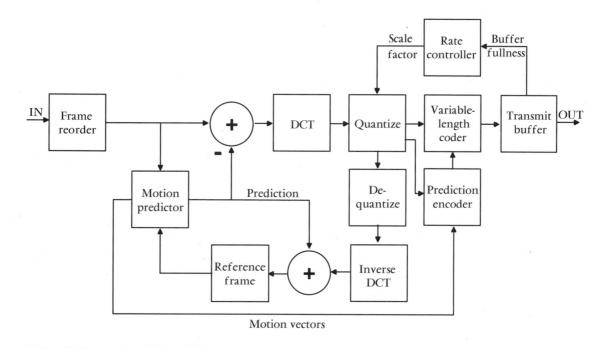

Figure 9-9 Simplified MPEG encoder.

In the decoder, spatial data is decoded as with JPEG. The output will, depending on the coding method employed, be decoded spatial data, or information from the reference frame, or—when residuals are transmitted—a combination of the two.

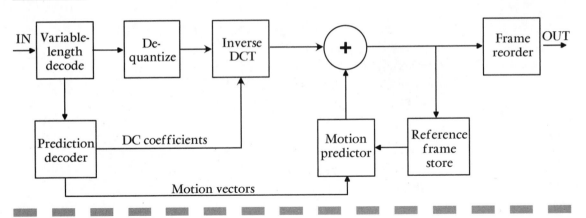

Figure 9-10 Simplified MPEG decoder.

MPEG-2

Introduction

MPEG-1 was frozen (i.e., subsequent changes were allowed to be editorial only) in 1991. In the same year, the MPEG-2 process was started, and MPEG-2 eventually became a standard in 1995. The initial goals were simple. Because MPEG-1 had no tools to accommodate interlaced video and, as specified, supported only CIF quality at 25 or 30 frames per second, there was a need for a standard that would include broadcast quality video.[1]

In many ways, MPEG-2 represents the "coming of age" of MPEG. The greater flexibility of MPEG-2, combined with the increased availability of large-scale integrated circuits, meant that MPEG-2 could be used in a vast number of applications. It is important, however, not to belittle MPEG-1.

MPEG-1 is vastly more complex than JPEG, both in specification and implementation. Nevertheless, it is fair to look at MPEG-1 and characterize it as slightly modified JPEG plus temporal compression and rate control. The previous chapter looked almost exclusively at temporal compression and rate control because a knowledge of JPEG (gained from Chapter 7) provides almost all that is needed to understand the spatial coding and associated syntax of MPEG-1. To be sure, there are improvements to JPEG, but the changes are minor and based upon experience with an excellent system. This was not a given—at the beginning of MPEG development, many alternative spatial compression algorithms were tried, but the JPEG approach proved superior.

A similar commentary can be made about MPEG-2. It is much more complex than MPEG-1: the standards documents outweigh MPEG-1 documents to a considerable degree, and for the same sample rate, an MPEG-2 encoder is about 50 percent more complex. At the beginning of the MPEG-2 process, 39 algorithms competed in subjective tests, some very different from MPEG-1. However, it is possible to summarize MPEG-2 quite simply—it is MPEG-1 with some improvements, plus tools for interlace, plus scalable syntax, plus a range of profiles and levels to accommodate a wide range of applications, plus a system layer to handle multiple program streams. The remarkable points are that the extensions are so versatile, and that the MPEG-1 base was such a robust platform on which to build.

---

[1] The "as specified" needs emphasis. The MPEG-1 syntax and techniques had proved to be robust and versatile well beyond the parameters of the Standard. Satellite broadcasting began using MPEG-1 syntax, and one of the proponents in the competitive phase of the U.S. advanced television development used MPEG-1 syntax for high-definition television.

Perhaps the success of MPEG-2 is best highlighted by the demise of MPEG-3. This exercise was started with the objective of providing a compression system suitable for high-definition television. It was soon abandoned when it became apparent that the versatility of MPEG-2 embraced this application with ease. MPEG-2 is, of course, the basis for the Advanced Television Systems Committee (ATSC) Digital Television Standard, and of the European Digital Broadcasting (DVB) now being implemented for both standard- and high-definition transmissions around the world.

MPEG-2 Enhancements

Color Space

All profiles and levels of MPEG-2 support 4:2:0 coding, but this is subtly different from MPEG-1 in that the position of the color difference samples is redefined.

Some profiles and levels support 4:2:2 and 4:4:4 coding. The corresponding macroblock structures, and the positioning of color frame samples, are shown in Figure 10-1.

Slice Structure

As a reminder, a *slice* is a collection of macroblocks, in scan order. This is significant because a slice may be decoded without reference to any other slice; for example, all predictive coding is reset at the first macroblock of a slice.

MPEG-1 has no restrictions on slice size—a slice may be a single macroblock, or the entire picture, or anywhere in between. In contrast, MPEG-2 requires that a slice be contained within a single row of macroblocks. The slice may be a complete row of macroblocks, or less, but never more.

Quantization

MPEG-1 allows only 8-bit precision for the DC coefficients of DCT blocks. Some MPEG-2 profiles and levels permit 9- or 10-bit precision to be specified for these coefficients.

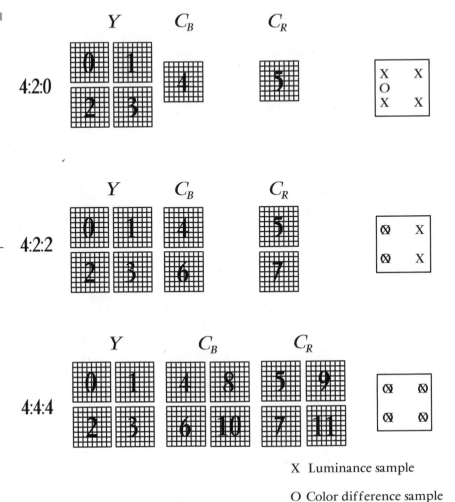

X Luminance sample

O Color difference sample

Concealment Motion Vectors

The concealment motion vectors tool is intended for use when the MPEG bit stream is transmitted over a lossy channel, and there is a significant probability that some macroblocks will be lost. This is particularly critical for intra coded blocks in I- and P-frames, because these errors may propagate over many frames.

MPEG-2 allows for concealment motion vectors (CMVs) to be transmitted with intracoded macroblocks. If a macroblock is lost, the CMV in the macroblock immediately above will point to a similar macroblock, which may be substituted in the decoder.

3:2 Pulldown

In 60-Hz countries (using television systems with 30 frames per second), 24-Hz motion picture film is transmitted (in conventional systems) by alternately using two fields then three fields for a film frame. This sequence repeats indefinitely (five fields = two film frames = one-twelfth second). This is inefficient coding by conventional MPEG techniques. Three pictures out of five carry complete representations of a film frame, but the other two out of five carry two fields from different film frames. These may be difficult to encode (using many bits), and the information they carry is totally redundant—each of the fields is already carried by another picture as part of a complete frame.

MPEG-2 permits the 3:2 sequence to be extracted at the encoder and, in this mode, 24 frame pictures are sent each second instead of 30. Each of the 24-Hz pictures represents a complete film frame. If the receiver has a 60-Hz display, the decoder is instructed to recreate the 3:2 sequence by displaying one picture for three fields, the next for two fields, and so on. This represents a substantial improvement in efficiency.

Pan and Scan

MPEG-2 can support displays of different *aspect ratios.* For example, a program may be transmitted with a 16:9 aspect ratio. *Pan and scan* information may be associated with the pictures to instruct the decoder which part of the image to display on a 4:3 display.

The mechanism supports both horizontal and vertical offsets (so that it could be used to display 4:3 material on a 16:9 display). The offsets are specified in increments of one-sixteenth of a pixel, and can be sent every field, so that smooth pans are quite practical.

MPEG-2 Profiles and Levels

Chapter 9 described the concept of MPEG compliance. In particular, it was stressed that a compliant decoder had to be capable of decoding any valid MPEG-1 bit stream. With the constraints of the MPEG-1 parameters, this was not a major issue. However, during the development of MPEG-2, it became apparent that the same rule could not be applied across all possible implementations of MPEG-2. A decoder that could support all the new syntax elements at all permissible data rates would be grossly uneconomical for simpler applications. The committee realized that unless a better way was built into the standard, a variety of incompatible proprietary subsets would be built. The solution was to define a range of sensible subsets within the standard. It turned out that divisions were required in two directions.

MPEG-2 consists of all the MPEG-1 coding tools, plus a considerable number of new ones. These include the tools for coding interlaced pictures and an array of tools for generating several different types of scalable bit streams (streams from which a base layer may be extracted and, as an option, additional layers representing picture enhancements or additional information). The tools, or more accurately, the syntactical elements that may be used are determined by the MPEG-2 *profile*. MPEG-1 bit streams are accommodated for backward compatibility. An even simpler profile is provided that does not permit B-frames, allowing considerable simplification and reduction in memory requirements at the decoder.

Most of today's MPEG-2 work is performed at *Main Profile*. The principal differences in syntax elements between MPEG-1 and Main Profile MPEG-2 are the provision of tools for interlace and the ability to scale to large pictures and data rates.

Three profiles provide bit streams with scalable elements. These are discussed in below.

A more recent addition is the *4:2:2 profile*. This was added at the request of the professional television community, mainly for use in studios. Main Profile met all the requirements for tools for this application, but supported only 4:2:0 coding. The 4:2:2 profile is defined by MPEG at Main Level (see below), but its extension to High Level is being standardized by SMPTE for "lightly compressed" high-definition signals in studio applications, and for program contribution and distribution links.

Profiles define the tools or syntactical elements that may be used; *levels* define the permissible ranges of parameters within the bit stream. Again, the various levels are provided to allow practical subsets to be implemented in a standard manner and with a standard nomenclature. As pointed out in the MPEG standard, it is possible to encode pictures "as large as (approximately) 2^{14} pixels wide and 2^{14} lines high." Clearly it would be ridiculous to specify that any compliant decoder be capable of handling images of this size.

Four levels are defined, *Low, Main, High-1440* (intended for the anticipated European high-definition system with 1440 pixels by 1152 active lines), and *High*. Not all profiles are defined at all levels. A combination of level and profile is specified as (for example) "Main Profile at Main Level" or as MP@ML.

TABLE 10-1　Profile/level pairings for MPEG-2

Level	Profile							
	Simple	Main	SNR	Spatial		High		4:2:2
				Enhancement layer	Base layer	Enhancement layer	Base layer	
High		1920H 1152V 60 Hz				1920H 1152V 60 Hz	960H 576V 30 Hz	1920H 1152V 60 Hz
High 1440		1440H 1152V 60 Hz		1440H 1152V 60 Hz	720H 576V 30 Hz	1440H 1152V 60 Hz	720H 576V 30 Hz	
Main	720H 576V 30 Hz	720H 576V 30 Hz	720H 576V 30 Hz			720H 576V 30 Hz	352H 288V 30 Hz	720H 512/608V 30 Hz
Low		352H 288V 30 Hz	352H 288V 30 Hz					

Table 10-1 provides a partial matrix of profile/level definitions, showing which combinations are defined and a few of their parameters. Note that 4:2:2@HL is not documented by MPEG, but is the subject of a standard prepared by the Society of Motion Picture and Television Engineers (SMPTE). Table 10-2 shows a greater set of parameter constraints for Main Profile at various levels. (The full set of constraints occupies many pages of the MPEG-2 standard.)

The tables show the maximum permitted value of each parameter. A compliant decoder is required to decode any compliant bit stream at its specified profile/level, and compliant bit streams at any lower profile and/or level.

Level	Parameter	Constraint
High MP@HL	Samples/line Lines/frame Frames/s Luminance rate Bit rate VBV buffer size	1920 1152 60 62,668,800 samples/s 80 Mbits/s 9,781,248 bits
High-1440 MP@H14	Samples/line Lines/frame Frames/s Luminance rate Bit rate VBV buffer size	1440 1152 60 47,001,600 samples/s 60 Mbits/s 7,340,032 bits
Main MP@HL	Samples/line Lines/frame Frames/s Luminance rate Bit rate VBV buffer size	720 576 30 10,368,000 samples/s 15 Mbits/s 1,835,008 bits
Low MP@LL	Samples/line Lines/frame Frames/s Luminance rate Bit rate VBV buffer size	352 288 30 3,041,280 samples/s 4 Mbits/s 475,136 bits

Interlace Tools

Interlace is a scanning system devised in the early days of television to increase the display flicker rate (so that flicker is virtually unnoticeable) without increasing the bandwidth of the television signal. The technique is to scan alternate lines of a frame in half a frame interval (called a field) followed by the remaining lines during the second field. In a 525-line system, for example, the complete set of 525 lines is scanned in one-thirtieth of a second, but the picture on the display is painted 60 times a second, an acceptable flicker rate. Of course, this does not come free. The technique overlaps parts of the vertical and temporal spectra, resulting in some artifacts when horizontal or near-horizontal edges move vertically in the picture.

Because of spectrum overlap, converting an interlaced picture to progressive format (*deinterlacing*) is not a trivial undertaking if there is motion in the picture. It can be accomplished only by techniques, such as motion estimation, similar to those described in Chapter 8.

MPEG-1 makes no concession to interlace. This does not mean interlace cannot be encoded, but the two interlaced fields must be combined into one frame that includes all the vertical lines. Of course, this frame now contains two sets of lines that represent different times. If there is vertical motion, the DCT blocks may contain alternate lines of pixels that are radically different (because they come from different fields), and coding may be very inefficient.

Many observers thought that when the proponents in the U.S. advanced television proceedings changed from analog systems to digital systems the death knell of interlace had been sounded. Many people "proved" that any possible advantage of interlace was more than offset by the penalty of encoding an artifacted image. The computer industry (which abhors interlace) rejoiced.[2]

Fortunately or unfortunately, depending on your point of view, they reckoned without the genius of the MPEG-2 workers. MPEG-2 offers a number of tools for coding interlaced pictures and the combination is extremely efficient. Viewing tests at the Advanced Television Evaluation Laboratory (ATEL) in Ottawa, Canada, with nonexpert viewers, showed that, for a given number of bits, interlace permitted a subjectively "better" picture than progressive scan. So the digital television system still carries the benefits/curse (delete as desired) of interlace.[3]

Frame and Field Pictures

MPEG-2, like MPEG-1, requires pictures, or frames, that include all the scan lines of the image. As with MPEG-1, these frames can be I-, P-, or B-frames. However, MPEG-2 permits many new variants.

---

[2] Computer applications like technical drawing programs use elements such as single-pixel lines that violate Nyquist in the spatial domain. Even on a progressive display, these artifact heavily when tilted slightly off the horizontal. This is regarded as an acceptable tradeoff that permits high-precision work with reasonable cost displays. If the display is interlaced, this artifact flashes at half the display rate, making the image essentially unusable.

[3] It must be added that the progressive system suffered from the lack of any equipment that could generate high-quality progressive pictures. In fact, the situation was so bad that for the second round of testing the progressive test sequences were derived from interlaced sources! There is no doubt that, even with the tools of MPEG-2, progressive sequences may be coded more efficiently than interlaced sequences; arguments continue as to the degree of benefit.

A coded I-frame can comprise a complete I-frame picture (as in MPEG-1), or a pair of I-fields, or an I-field followed by a P-field predicted from the I-field.

A coded P-frame can be either a P-frame picture or a pair of P-fields, and a coded B-frame comprises a B-frame picture or a pair of B-fields. The type of coding is indicated in the picture header, and may be chosen as required for each picture.

In a field picture, the fields are processed sequentially, so macroblocks contain only samples from a single field and represent an area of the image 32 lines high.

Frame and Field DCT

MPEG defines the two fields of a frame as upper and lower fields as shown in Figure 10-2. This distinction is used both in DCT coding and in predictions.

Figure 10-2
Field and frame pictures, and upper and lower fields in MPEG-2.

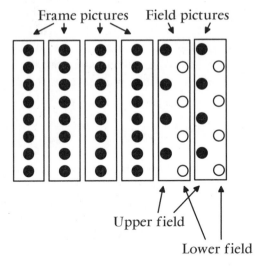

Frame pictures   Field pictures

Upper field

Lower field

MPEG-2 defines two types of DCT coding for a macroblock, *frame DCT* and *field DCT*. Frame DCT is exactly the same as seen in MPEG-1. The 16×16 block of luminance pixels is divided into four blocks of 8×8 according to placement. Field DCT uses the same horizontal split, but vertically it takes the eight lines from the upper field for the upper blocks, and the eight lines from the lower field for the lower blocks, as shown in Figure 10-3. The color difference blocks (which have only eight lines) are always assumed to belong to the upper field. Field DCT is more efficient when there is a significant difference between the two fields, usually as a result of motion.

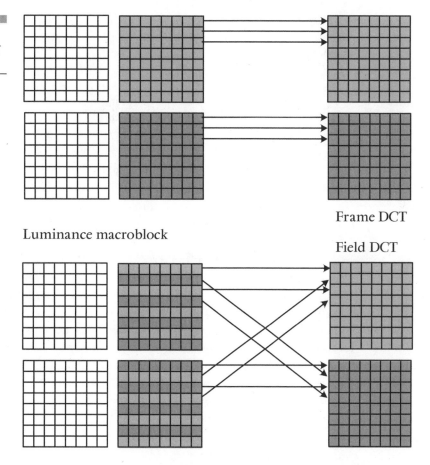

Figure 10-3
Field and frame DCT in MPEG-2.

Luminance macroblock

Frame DCT

Field DCT

Frames coded as two fields (field type) always use field DCT, because that is how the macroblocks are formed. Frame pictures may use frame DCT or field DCT, selected at the macroblock level (and each macroblock header must specify the DCT type used). Generally a frame with overall vertical motion is best coded as a field picture; a frame with only small areas of vertical motion may be best coded as a frame picture.

Another tool available to the encoder is an alternative scan pattern for the quantized AC coefficients, shown in Figure 10-4. This scanning system is designed to maximize zero-run-lengths in the presence of vertical energy resulting from motion in interlaced pictures.

Figure 10-4
Alternate coefficient scan for interlaced pictures.

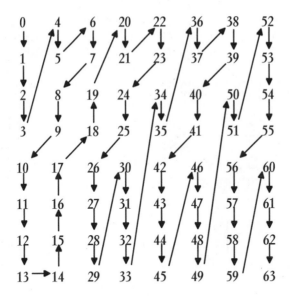

Frame and Field Prediction

Frame prediction is described in the previous chapter on MPEG-1. The motion estimator looks for the best 16×16 region in the reference frame to match the luminance samples of the macroblock being coded.

Field prediction separates upper and lower fields in the macroblock being encoded and in the reference frame. The search is for a region that matches the 16×8 array of pixels representing the upper or lower field of the current macroblock.

As may be expected by now, the rules are different for frame pictures and field pictures. In frame pictures, either frame prediction or field prediction may be chosen at the macroblock level. If field prediction is chosen, a motion vector is determined for the upper field of the macroblock by the best match found in either the upper or lower field of the reference picture. Similarly, a motion vector is found for the lower field of the macroblock, again searching in both fields of the reference frame. However, only the better of the two matches is used, and that motion vector is used to predict both the upper and lower fields of the macroblock.

In field pictures, field prediction looks for a match for the 16×16 samples (all from the same field) in the two previous fields. For the upper field (processed first) the two previous fields are the two fields of the previous picture. When the lower field is processed, the two previous fields are the lower field of the previous picture, and the upper field of the current picture. In each case, the chosen motion vector applies to the complete macroblock.

Field pictures may also use 16×8 motion compensation. In this mode, separate motion vectors are derived from the two previous fields for the upper and lower 16×8 sections of the macroblock. Both motion vectors are transmitted, and each is used for the corresponding part of the macroblock. Previous fields are determined as described in the preceding paragraph.

A final tool for encoding interlaced pictures of either frame or field type is *dual-prime prediction*. In this mode a single motion vector is found for a 16×8 single-field region of a frame picture (16×16 region of a field picture) by searching in the previous field of the same parity (upper or lower). A second incremental vector (value −1, 0, or +1 in both x and y) is derived by finding the best match (from the nine possibilities) for the same region, but in the field of opposite polarity. At the decoder, the two blocks identified by the vector and the incremented vector are averaged to form the predictor block. This technique has proved to be especially efficient.

Scalable Coding Profiles

As mentioned earlier, several MPEG-2 profiles permit some form of scaling. In each case the bit stream is divided into a base layer, which must always be decoded, plus one or more enhancement layers where decoding is optional. These profiles are not much used today, but are expected to be important for the future. It should be noted that dividing information into two or more layers for encoding carries a penalty in efficiency; it is more efficient to code all the information in a single layer.

In some environments, such as broadcasting, the base layer could be modulated at greater depth than the enhancement layer. Receivers with a good path to the transmitter could decode all of the signal and receive the benefits of the enhancement layer; receivers on the fringe of the transmission would be able to decode only the base layer (a poor signal-to-noise ratio [SNR] would prevent decoding of the enhancement layer).

In other applications the techniques can be used for backward compatibility. An "old" decoder would not recognize the enhancement layer and would not attempt to decode it, but would receive the base layer as designed. A "new" decoder could decode all layers and provide the enhanced output.

More commercially, a video-on-demand service might deliver the base layer for one price, and the enhancements for a supplemental charge.

SNR Scalable Profile permits a low-rate, low signal-to-noise bit stream to be supplemented by additional information that, if used, improves the signal-to-noise ratio of the image.

The *Spatially Scalable Profile* permits decoding of a base level image by a simple decoder, and provides an enhancement layer that permits a more complex decoder to recover a higher-resolution image. There is an optional temporal scalability syntax that permits a similar operation with different temporal resolutions. The *High-Level Profile* permits several enhancement layers and virtually all combinations.

It is speculated that one of these mechanisms may be used in the future to allow the U.S. ATSC system to extend to 1920×1080/60 progressive. With spatial scalability, the base layer might be 1280×720/60p; with temporal scalability, the base layer could be 1920×1080/60i (interlaced). In either case, the base layer only would be decoded by a first-generation receiver, which would ignore the enhancement layer. A later-generation receiver would decode both layers to provide the higher-quality pictures.

MPEG-2 System Layer

In the real world, delivery of useful program material requires audio as well as video, and may require additional data streams such as captioning. Clearly, it is not a good idea to send all the video, then start the audio, so the bit streams must be *multiplexed* together. In some situations we may wish to send several different programs, each with its associated video, audio, etc., in the same bit stream. Because we are sending continuous information (such as audio) in a discontinuous manner (packets), there is a need for timing and control to ensure that all the inputs can be reassembled correctly.

All this work is the job of the systems layer of an MPEG standard. MPEG-1 had a very simple systems layer designed to work with digital storage media such as CD-ROMs, which are close to error free. It could multiplex several videos and audios into a stream, but only if they had a common timebase.

The requirements for MPEG-2 were more diverse. It had to work over ATM networks and other lossy transmission systems, so there was a need for greater error resilience. There was a need to handle multiple programs without requiring a common timebase, and backward compatibility with MPEG-1 was also required.

MPEG-2 systems define two bit stream constructs. The *program stream* (PS) is modeled on MPEG-1, and compliant PS decoders will decode MPEG-1 bit streams. The program stream requires that all components share a common timebase and is designed for low-error environments. The alternative is the MPEG-2 *transport stream* (TS). This is a more robust mechanism, capable of carrying multiple programs without the need for a common timebase. The transport stream is designed for use in error-prone environments.

Before looking at the PS and TS in detail, we need to consider the components that are common to both.

Packetized Elementary Stream

The output of a video encoder, or an audio encoder, is known as an *elementary stream*. If the stream is divided into packets for transmission to the multiplexer, it is known as a *packetized elementary stream* (PES). PES packets are of variable length, depending on how the information is assembled in the particular encoder. For example, a video coder might make each complete picture into a packet. Each packet starts with a header that includes identification of the payload, timing information, etc.

Program Stream

The program stream is the simpler of the two communications mechanisms provided by the MPEG-2 systems layer. As mentioned above, it is intended for low-error environments and makes no special provision for error handling.

The program stream consists of packs, each containing one or more PES packets (Figure 10-5). The pack header contains synchronization information. The program stream can accommodate up to 16 video and 32 audio streams, but in MPEG parlance these are regarded as parts of a single program because they must share a common timebase. Every pack is stamped with information from a single *system clock reference* (SCR).

Figure 10-5
MPEG-2 program
stream.

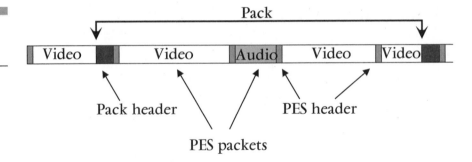

Transport Stream

The transport stream is a much more sophisticated communications system capable of carrying several programs, each made up of several elementary streams. All the streams within a single program must share a common timebase, but each program can use a separate time reference.

In a noisy transmission channel, there are substantial advantages to the use of relatively short, fixed-length packets. These packets are particularly suitable for protection by forward error correction. Redundant data is added to the packet, generated by a known algorithm, such as Reed-Solomon. When the data is examined at the receiver, the redundant data allows detection of errors and correction of some number of errors per packet.

MPEG does not specify error correction; indeed it would be wrong to do so because error correction must be chosen according to the error characteristics of the particular communication channel. There is no universal remedy. The transport stream facilitates the subsequent addition of error correction by arranging all transmitted data in fixed-length packets, 188 bytes long. (Compare this to the program stream, where packets of 1 to 2 kbytes are typical, and lengths of up to 64 kbytes are permitted.)

Each TS packet has a 4-byte header that includes a *packet identification* code (PID). All packets carrying the same service (e.g., program no.1 video) have the same PID, and a sequence number in the packet header ensures that the packets are decoded in the correct order. The other 184 bytes may be payload, or some combination of adaptation field and payload (Figure 10-6). The adaptation field is used when additional control information is required, and its presence is indicated in the packet header.

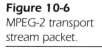

Figure 10-6
MPEG-2 transport
stream packet.

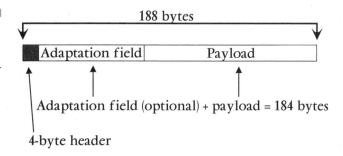

One control mechanism that uses the adaptation field is the *program clock reference* (PCR). This is the timing mechanism for the communications channel. PCR is a 27-MHz reference signal generated in the multiplexer. At least 10 times a second, a sample of this clock is sent, via the adaptation header, to the demultiplexer. The demultiplexer maintains its master clock, synchronized to the multiplexer clock by means of the PCR samples. This is the fundamental mechanism for timing control in the transport stream.

Locked to the 27-MHz PCR is a 90-kHz clock that generates time values for the rest of the system. Consider a PES packet from the video encoder, representing one picture, and described by the PES header. When this PES packet is split into transport packets, the PES header always immediately follows the (4-byte) transport header. To this is added a *presentation time stamp* (PTS), generated from the 90-kHz clock. Audio/video synchronization is achieved by sending audio and video with the same PES to the audio and video display systems at the same time.

Because some data (like video anchor frames) is deliberately sent out of order, a second time stamp, also a value of the 90-kHz clock, called a *decode time stamp* (DTS) is added where required. The DTS is used to control ordering of data into the decoder.

Given that every stream has a unique PID, separation of the data into streams appropriate for different decoders is easy, provided we know the PIDs. The transport stream provides a number of tables, linked to the multiplex structure, to convey this information—see the multiplex shown in Figure 10-7.

Audio, video, and data PES are multiplexed into Program 1, together with a *program map table* (PMT). The PMT is a TS packet that contains as payload the PIDs for each of the elementary streams of Program 1. A different set of audio, video, and data is multiplexed with another PMT to form Program 2, and so on. The various programs are multiplexed with a *program association table* (PAT) to form the trans-

Figure 10-7

MPEG-2 transport
stream multiplexing.

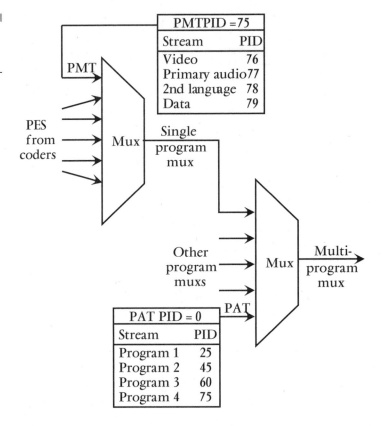

port stream. The PAT, which is also a TS packet, carries as payload the PIDs that identify the various PMTs. The final key to decoding is that the PAT *always* has PID = 0.

So, to find, say, the primary audio of Program 2 we first look for a packet with PID = 0; this is the PAT. Within the payload of the PAT we will find the PID allocated to the PMT of Program 2. We then look for a packet with that PID, and as its payload we find the full list of PIDs that identify the various elementary streams of Program 2.

MPEG-2 also permits other table types within the transport stream. These include network information, conditional access, and private tables. Their use is beyond the scope of this book.

AATSC has published a standard (A/65) for Program and System Information Protocol (PSIP). This provides a user-friendly mechanism for identifying and selecting among the various program services in a broadcast multiplex, and may also include electronic program guide information. Of particular interest here, PSIP can provide PID information for each of the streams within a program service. This speeds the tuning process when channel surfing because the decoder does not have to go through the PAT, PMT sequence and, more importantly, does not have to wait for each of these packets when a channel change command is given.

Practicing the Art of MPEG

This section includes material adapted from a presentation by Dr. Michael Isnardi, with the permission of the Sarnoff Corporation.

Contributors to Poor Performance

As has been stressed throughout this book, many assumptions are made in the design of a compression system. Attempting to encode any content that violates these assumptions will almost certainly have an adverse effect on performance. Two results are likely. The nonconforming material will probably be distorted by the encode/decode process. Often more significantly, the compression system will probably use a disproportionate number of bits to encode the nonconforming content, with a resulting degradation in performance on normal content.

The most obvious example for any video compression system is aliased image content. It cannot be stated too strongly that the best starting point is a clean image, free from sampling artifacts. The best starting point for this is, of course, a clean analog signal, accurately Nyquist filtered, and as free as possible from noise components. In this context, noise components include random noise, film grain, scratches, etc. As discussed earlier in the book, sophisticated preconditioning units are available that incorporate many different types of filter.

The nature of the image itself will have a substantial effect on the occurrence and the visibility of artifacts. Soft, static images show artifacts clearly, but these images are easy to encode, so there should be very few artifacts. Highly detailed images generate more artifacts, but are also quite good at masking them. Motion, particularly complex motion, makes the scene more difficult to encode and is likely to result in artifacts, but this does not mean that a scene full of motion is a good test. Moving objects should be soft (blurred), and unless the eye is tracking a moving object, motion is a very effective mask for most artifacts.

The worst type of sequence is one that contains considerable static detail in some areas, and/or complex motion in other areas, plus areas with gentle gradients where the results of bit starvation may readily be seen. Note, however, that a realistic test sequence must be fair to be useful. Any compression system can be broken by the introduction of a sufficiently bad signal, such as one containing large areas of noise, and little is proved by this exercise.

Other types of picture content that perform contrary to the design assumptions will cause reductions in efficiency. MPEG assumes a linear translation model for motion and handles this well. Other types of image motion such as rotations and zooms will not be predicted efficiently and will load the system heavily. Similarly, transparent or translucent moving objects and dissolves including moving objects are sources of trouble.

MPEG has no effective mechanism to track brightness changes, so changing light, shadows, and fades all result in poor prediction. Theory would suggest that a GOP structure with a single B-frame between each pair of anchor frames would cope better with these image components, using the interpolation mode, but I do not know whether this has ever been tried on real images. Anyway, such GOPs are not very efficient for other content.

Uncovering regions of high detail will result in frames needing a great deal of intracoding, and consequent bit starvation. Scene changes have a similar effect, but the resulting artifacts are often masked by the temporal effects of the scene change itself in the human visual system.

MPEG Artifacts

Blocking artifacts are a fundamental characteristic of stressed DCT systems. If for any reason there are insufficient bits available, the block and/or macroblock structures become visible. This type of error is most evident when the image area is relatively featureless, and particularly when there is a gentle gradient. Even if individual blocks are not discernible, the human visual system is very sensitive to small correlated brightness steps, and often a rectilinear boundary will be visible in scene elements such as plain concrete or pavement surfaces.

Blocking may also be particularly noticeable when the eye tracks a fast-moving object; in this instance the effect is apparent because the object moves over the stationary block pattern.

Mosquito noise is a characteristic of all quantized DCT systems and appears on sharp edges in the scene, such as titles. The edges generate coefficients right across the block (for vertical edges) and the high-frequency coefficients are quantized more coarsely than those for the lower frequencies. This has the effect of spreading the energy spatially when it is transformed back into the pixel domain, producing a characteristic patterning along the original edge.

Dirty window artifacts appear as streaks or noise that remains stationary while objects appear to move behind—rather like viewing the scene through a dirty window. This is usually caused by insufficient bits being allocated to code motion prediction residuals.

Wavy noise is often seen during slow pans across a very detailed scene, such as a crowd shot at a sporting event. Like mosquito noise, this is a result of coarse quantization of the high-frequency coefficients, but the motion causes the spreading to vary periodically as the detail moves across the DCT blocks.

Tips for Higher Quality

At the risk of being repetitive, tip number one is: make sure the video being coded is clean. A good preconditioning unit, supplementing any preconditioning in the encoder itself, can substantially improve coder performance. However, it is important not to replace one artifact with another. In many demonstrations designed to prove that MPEG-1 can produce reasonable pictures over a T1 link at 1.5 Mbits/s, the desire to eliminate blocking led people to introduce ridiculous degrees of temporal filtering. If characters disappear from one side of the screen and reappear gradually on the other, you have wound up the control too high!

Also a familiar refrain: do not encode more than is necessary. Do not encode the picture edges if these contain only blanking edges and extraneous content (this is why standard MPEG encodes only 704 of the 720 pixels in the 601 standard). How much resolution is needed? In a busy scene, static resolution may not be important to the viewer; better results may be obtained by reducing the horizontal resolution to free up more bits to handle motion.

Much improvement will come from improved encoder designs. New techniques for motion estimation, such as phase correlation, will result in better correlation of motion vectors and more efficient coding. Better rate-control algorithms will ensure that each bit is used where it will do most good. Efficient statistical multiplexing algorithms (see Chapter 19) can help by maintaining consistent quality—often perceptually more important than absolute quality.

MPEG-4

Introduction

The wheels of international standardization grind slowly, and to ensure that a standard is eventually achieved there are strict rules that prohibit substantive change after a certain point in the process. By the time a standard is officially adopted there is often a backlog of desired enhancements and extensions. So it was with MPEG-2. As discussed in Chapter 10, MPEG-3 had been started and abandoned, so the next project became MPEG-4.

MPEG-4 exists in two parts, Version 1 and Version 2, which is a superset of Version 1. In other words, all Version-1 constructs are valid within Version 2. Work began in 1993, and was formalized in July 1995 with a call for proposals. The various proposals for audio and video were evaluated by subjective testing and by experts. Drafting of the Version 1[1] followed, leading to a Committee Draft in November 1997, and became an ISO/IEC International Standard in 1999. Meanwhile, work continued on the Version-2 extensions and the Draft Amendment was frozen in December 1999, with various phases of formal adoption during 2000. As of the summer of 2000, work on extensions continues forming the basis for Amendments 2 and 3, representing Versions 3 and 4 respectively. Even more work is planned, to determine and meet the needs of digital cinema.

At first, the main focus of MPEG-4 was the encoding of video and audio at very low rates. In fact, the standard was explicitly optimized for three bit rate ranges:

- Below 64 kbits/s
- 64 to 384 kbits/s
- 384 kbits/s to 4 Mbits/s

Performance at low bit rates remained a major objective and, as we shall see, some very creative ideas contributed to this end. Great attention was also paid to error resilience, making MPEG-4 particularly suitable for use in error-prone environments, such as transmission to personal handheld devices. Other profiles and levels use bit rates up to 38.4 Mbits/s, and work proceeds on studio-quality profiles and levels using data rates up to 1.2 Gbits/s.

More important, MPEG-4 became vastly more than just another compression system—it evolved into a totally new concept of multi-

---

[1] Except for Conformance Testing, which followed about a year later.

media encoding with powerful tools for interactivity and a vast range of applications. Even the official "overview" of this Standard spans 67 pages, so in this chapter I can only attempt a brief introduction to the system. I hope I will manage to convey some of the new concepts and demonstrate the enormous richness of MPEG-4; it really is a Standard with something for everyone, and each facet of the Standard offers a wealth of tools and options. Despite the focus on the low bit rate applications, MPEG-4 includes important extensions for high-end studio operations and high-definition television.

The most significant departure from conventional transmission systems is the concept of *objects*. Different parts of the final scene can be coded and transmitted separately as *video objects* and *audio objects* to be brought together, or *composited*, by the decoder. This raises a number of interesting possibilities. Most obviously, different types of objects may be coded using different techniques, and each may be coded with the tools most appropriate to the job. The different objects may be generated independently, or in some cases a scene may be analyzed to separate, for example, foreground and background objects. An interesting example was demonstrated recently. Video coverage of a soccer game was processed to separate the ball from the rest of the scene. The background (the scene without the ball) was transmitted as a "teaser" to attract a pay-per-view audience. Anyone could see the game, but only those who paid could see the ball! Figure 11-1 shows the general concept.

As we have seen throughout this book, no coding system is universally efficient. DCT and quantization are effective for continuous-tone, bandwidth-limited images, but certainly not optimal for everything that we may wish to display. As a simple example, if a scene is to be overlaid with text, conventional systems key the text into the video, often filling the body and edges of the letters with contrasting colors. The resulting combination is then coded by normal means. Superimposing text in this manner adds a good deal of high-frequency energy to the signal, and these edges are not well coded by DCT. They consume a disproportionate number of bits, and the result is often less than pleasing. One of the very noticeable artifacts of DCT-based encoding is the effect known as *mosquito noise* on sharp edges such as overlaid text, caused by excessive quantization of the high-frequency coefficients.

Thus the addition of text has a significant impact on the efficiency of the video coding. Yet we know that we could code that text in a very simple manner—the letters can be specified in ASCII (or some more efficient code), and position, font, size, color, and spacing information

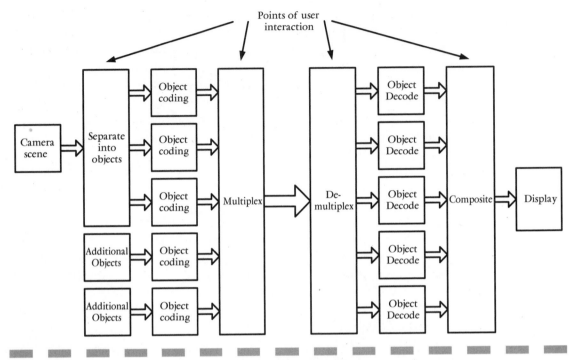

Figure 11.1 Objects and object coding in the MPEG-4 video system.

can be added with a fairly small number of bits. But for this to work, the decoder needs the ability to generate the titles from the information supplied, and to key the titles over the decoded video prior to display. Decoding of MPEG-4 video bitstreams may require a decoder with multiple decode engines and the ability to perform compositing operations.

MPEG-4 is not confined to video and text. We discuss some of the other options later, but for the moment let's continue with the text example. The ability to composite elements at the decoder provides far more than coding efficiency. In conventional systems, the image to be displayed is determined and constructed prior to transmission; overlaid titles become a part of the video prior to encoding and transmission. In MPEG-4, it is possible to transmit several text streams and to choose at the decoder which stream or streams (if any) to combine with the video. The choice may be entirely that of the viewer, or may be constrained by other information in the transmitted bit stream.

A simple and crude example of this capability exists in today's analog television system. "Closed captions" for the hearing impaired are coded into one line at the top of the picture, and the viewer may choose whether or not to overlay them onto the video. MPEG-4 offers vastly more flexibility and efficiency even with simple text; color, position, and font can all be specified by the author, and the viewer may have the choice, for example, of several languages.

We will look in more depth at the capabilities of the system, but we have already identified three key characteristics of MPEG-4 streams:

- Multiple objects may be encoded using different techniques, and composited at the decoder.
- Objects may be of natural origin, such as scenes from a camera, or synthetic, such as text.
- Instructions in the bitstream, and/or user choice, may enable several different presentations from the same bitstream.

These capabilities do not have to be used—MPEG-4 provides traditional coding of video and audio and improves on MPEG-2 by offering increased efficiency and excellent resilience to errors. However, the true power of MPEG-4 comes from the new applications made possible by the architecture described above. The independent coding of objects offers a number of advantages. Each object may be coded in the most efficient manner, which more closely emulates the processes of the human psychovisual system; the separation of objects allows interaction with these objects—particularly powerful in games and educational programs. Also, different spatial or temporal scaling may be used as appropriate. As an example, if bandwidth and/or computational resources are limited, it may be appropriate to retain full temporal resolution for an important foreground object, but to update to the background at a lower rate.

There are also a number of disadvantages to the object approach. The decoder must be capable of decoding all the possible bit streams it supports, and it must have a compositing capability. Thus a hardware MPEG-4 decoder is necessarily more complex than a comparable MPEG-2 decoder. Similarly, more code is necessary to implement a software decoder, but this decoder can have considerable flexibility in using the limited hardware resources to the best advantage.

Up to now, we have discussed only conventional video images and text. MPEG-4 provides other video objects, some of which are far more innovative; these are discussed in the next section. Clearly a proliferation of object types complicates decoder design, and it would be uneconomical to require all decoders to support all possible constructs and objects. Not surprisingly, MPEG-4 defines a number of profiles and levels, in a similar manner to MPEG-2, permitting implementation with different tool sets and various parameter limitations.

Video in MPEG-4

Before looking at the techniques used for video compression in MPEG-4 we must examine the structure of a video scene, as defined for MPEG-4. A typical scene might include a background, one or more foreground objects such as furniture, one or more people, and some graphic elements. In conventional video thinking, and in MPEG-1 and MPEG-2, the complete scene is sampled once per frame, producing a bitmap that is coded by the techniques discussed in earlier chapters. MPEG-4 can also work this way, but it can also handle each of the objects separately, if the information is presented in the correct way.

To make life simple, let's forget about the furniture for now; we will just assume it is a part of the background. So, apart from any graphic elements, the scene consists of a background and (say) one person as a foreground object. It is well known that the scene may not be that way in the studio—the person, such as a weather forecaster, may be standing in front of a plain blue or green background and *chroma keyed* over a different background (such as a weather map). In the studio, the image of the person in front of the plain-color background is processed to eliminate the background color and to generate a *key signal* or *alpha channel* representing the shape (and, perhaps, the transparency) of the foreground person. This shape information is used to composite the two scene elements together. Where the foreground person stands, the background scene is replaced by the image of the person. Elsewhere (outside the shape) the background is unchanged. If the person is holding a semi-transparent object like a glass of water, the final image must be a mix of foreground and background scenes, determined by transparency information contained in the key signal. For realism, a mix is also needed along the edges of the foreground shapes.

In MPEG-4 terminology, the person in the foreground is known as a *video object*, represented by two elements, the video image of the person, known as the *texture*, and the key signal or alpha channel, known as the *shape.*

The background image does not really need a shape associated with it, but MPEG-4 treats this as having a rectangular shape, a degenerate case.

MPEG-4 Video Hierarchy

The starting point for video in MPEG-4 is a visual concept rather than a signal, such as a foreground person in a scene. Before we can do anything with the object it must be sampled. Most objects are sampled at regular time intervals (frames), and each time sample (a complete spatial representation at one instant in time) is known as a *video object plane* (VOP). So, each object in the scene is represented by a series of video object planes—except a sprite or other static object, which may be represented by a single video object plane.

In more familiar terms, a camera views a scene and captures the information by sampling the scene (by scanning, or shuttering and scanning). The camera provides its output as a sequence of frames or, in MPEG-4 terminology, the texture part of a sequence of VOPs.

A VOP contains texture data and either rectangular shape information or more complex shape data associated with the object. VOPs, like frames in earlier versions of MPEG, may be coded with intradata, or by using motion compensation; this is discussed below.

Next up is the optional *group of video object planes* (GOV) that groups together video object planes. GOVs are similar to the groups of pictures (GOPs) in earlier MPEGs, and provide points in the bitstream ⸻ᵗᵘʳᵉˢ ⸻ 's are encoded independently from each other, and so pro- ⸻ccess points into the bitstream.

⸻*ct layer* (VOL) permits scalable coding of a sequence of ⸻. Scaling may be spatial or temporal and is discussed ⸻e VOLs correspond to multiple scalings of a sequence. ⸻matched to a set of available resources—usually limited ⸻d/or limited computational power.

⸻*bject* (VO) level includes everything in the bitstream ⸻cular video object. It includes all VOLs associated with ⸻ultiple video objects are represented by multiple VOs.

⸻*e video session* (VS) is the top video level of the MPEG-4 ⸻cludes all video objects, natural or synthetic, in the scene.

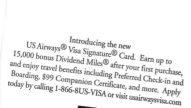

⸻oding

⸻o types of shape for video objects in MPEG-4, rectangular ⸻y. *Rectangular shapes* are trivial and merely indicate the extent of an image. Note, however, that this still represents a significant increase in flexibility over earlier standards. In MPEG-2, for example, the extent of the image is hard-coded into the bitstream headers. In MPEG-4, the rectangular size of a simple background video object is comparable, but there can still be other rectangular objects (of differing size) in the same session, such as picture-in-picture, and sprites.

Arbitrary shapes are the MPEG-4 equivalent of the graphics alpha channel. The shape again represents the extent of a video object, and at any point in the image plane it determines whether its associated object may be visible. The shape is defined over a rectangular area, known as the *mask*, sized to accommodate the greatest horizontal and vertical dimensions of the object itself. Both the horizontal and vertical sizes of the shape mask are multiples of 16 pixels.

Arbitrary shapes may be coded as binary data or grayscale data. Binary shapes are the simplest and indicate whether the object is transparent or opaque at any given point. The edge between the object and background is always hard. This normally represents a violation of Nyquist, and aliasing is likely to be visible. Grayscale shapes, on the other hand, permit a smooth, bandwidth-controlled transition or "blend" between objects and can yield much greater realism.

Binary shapes are used for simple applications and can represent total transparency or total opacity. In image areas where the video object is determined to be transparent, it cannot be seen under any circumstances. The background, or the composite formed by the background and any lower-level video objects, will be seen (unless occluded by some other higher-level object). Where the video object is opaque, the background and any lower-level objects will be invisible, and the video object associated with the shape will be visible, unless occluded by a higher-level object (something in front).

Binary shapes are coded into blocks 16 pixels square, like the macroblock used for textures, but they are known as *binary alpha blocks* (BABs). There are three classes of block in a binary mask; those where all pixels are transparent (not part of the video object); those where all pixels are opaque (part of the video object); and those where some pixels are transparent and others opaque. Coding of the first two cases is trivial. Blocks of the third type are those representing the boundary of the video object and are coded using techniques derived from arithmetic coding, discussed briefly in Chapter 3.

The algorithm used is known as *context-based arithmetic encoding* (CAE), extended to include the use of motion estimation. As in the conventional arithmetic encoding algorithm, coding is based upon a continually updated probability estimate. In the basic interCAE, the probability estimate is computed from ten pixels, above and to the left of the pixel being coded, see Figure 11-2a. For intraCAE, where motion estimation and prediction are used, the context includes some pixels from the current frame and some from the reference frame, as seen in Figure 11-2b.

Figure 11-2a
Context for intraCAE.

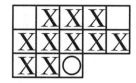

Figure 11-2b
Context for interCAE.

Reference VOP   Current VOP

Grayscale shapes are represented much as a luminance signal is, usually quantized to eight bits, with values from 0 (totally transparent) to 255 (totally opaque). The shape is coded using a philosophy similar to that used for binary shapes. Again there are macroblocks that are totally outside the boundary of the object, and those that are totally within the boundary of the object. Blocks outside are marked as "all zero" (fully transparent); those inside are marked as "all 255" (fully opaque). Blocks that are neither fully transparent nor fully opaque must be coded in a manner similar to that used for texture; that is, motion compensated DCT.

Texture Coding

Texture coding, the MPEG-4 term that corresponds to coding of conventional moving-image data, is built on MPEG-2 coding with extensions and improvements, as MPEG-2 was built on MPEG-1 and JPEG. In fact, a rectangular VOP is the closest equivalent to a "video frame" in pre—MPEG-4 terminology. The same concepts of intra- and intercoding apply, and video objects may be coded with I-, P-, and B-VOPs. All MPEG-4 profiles, except for the studio profiles described below, use a 4:2:0, *YUV* representation of video object textures.

In MPEG-4, not all video objects are the same size, and texture coding is necessary only for those areas that are part of the object. For rectangular objects this is simple; the size must be a multiple of 16 pixels (one macroblock) in each direction, and all macroblocks are processed. For objects with a more complex shape, the boundary is defined by the shape signal. The extent of the object is still defined by a rectangular array of macroblocks, but texture coding is performed only for those macroblocks wholly or partially inside the object boundary.

I-VOPs are coded much as MPEG-2 I-frames are, but several new techniques have been introduced to improve efficiency. In Chapter 4 we looked at predictive coding, and in Chapter 10 we saw that MPEG-2 used the simple $\check{X} = A$ predictor for encoding DC coefficients and motion vectors. Chapter 4 also showed that there are much better predictors, and that the performance of a simple adaptive predictor was substantially higher than that of other one- or two-dimensional predictors. MPEG-4 uses an adaptive predictor for the DC values. The predictor measures horizontal and vertical brightness gradients and predicts the DC value from the block above or the block to the left, in the direction of the lesser gradient.

Just as the correlation of the image provides benefit by predicting DC coefficients, it can also help in the coding of some AC coefficients. Areas of similar texture (in the conventional sense of the word) generate similar arrays of AC coefficients after the DCT transform. The similarity (and, therefore, the coding benefit) is greatest in the most significant coefficients; those that represent the greatest energy of the texture. These coefficients are normally the non-zero coefficients in the first row and/or first column. These are also the coefficients that will be quantized least aggressively and use the up most bits for transmission and thus offer the greatest potential for improved coding.

In MPEG-4, the AC coefficients of either the first row or the first column are predicted from those of the block immediately above, or immediately to the left, or above-left.

Quantization of the coefficients is similar to the method used by MPEG-2, but the mechanisms for coefficients scanning, and for variable-length encoding, are both improved.

The chosen method for coefficient readout is determined by the DC prediction. When there is no DC prediction, the zigzag scanning as described under MPEG-2 is used. If the DC coefficient was predicted from the block to the left, *alternate-vertical* scan is used, a scanning system that emphasizes reading out the vertical direction first. If, however, the DC coefficient was predicted from the block above, *alternate-horizontal* scan is selected, which results in a readout of coefficients with an emphasis on the horizontal direction first.

To improve the efficiency of variable-length encoding, two different VLC tables are provided, and the quantization level determines which table is used. The VLC codes themselves are reversible. This is part of the error-resilience strategy of MPEG-4; if there is an error in the received bit stream decoding can continue up to the error. Data after the error may be decoded by starting at the end of the block and decoding the VLC codes in reverse until the error is reached.

Boundary Coding

The ability to code objects of arbitrary shape creates an interesting situation at the object boundaries. Blocks that lie outside the boundary of the object need no texture coding. Those that are totally inside the boundary are coded normally according to the techniques we have discussed. Texture coding is needed for boundary blocks, but the object only exists for part of the block. If we chose to make the pixels outside the boundary of the object black, this would introduce a good deal of coefficient energy when the block was DCT transformed, yet this energy would not represent a meaningful part of the image. To avoid this inefficiency boundary, blocks are *padded*. First, all pixels that are not part of the object are given a value equal to the average value of all the pixels that are part of the object. The padding is refined by stepping through the pixels outside the object and performing a correction based on the average value of any neighbors that lie within the object. Note that no values within the object are changed. Changes to values outside the object do not affect the final result because these pixels will never be visible. The process described minimizes the energy of the coefficients when the block is DCT transformed.

Coding of Arbitrary-Shaped Video Objects

Now that we have looked at coding the different elements, it is worthwhile to examine all the possible cases for block coding within the mask of a video object.

In the simplest possible case, the block is flagged as transparent in the shape part of the VOP. This means that all the pixels in the block lie outside the boundary of the object, so that no texture coding is required.

The block may be flagged as opaque in the shape part of the VOP. In this case all pixels in the block lie inside the boundary of the object. No boundary coding is required, but texture coding is needed. The texture may be intracoded or predicted with motion compensation, as appropriate.

The remaining blocks are in some way part of the boundary area of the object. In the case of a binary shape this means there are some pixels that are transparent and some that are opaque. In the case of a grayscale shape, it means that not all pixels are transparent and not all pixels are opaque. Here both shape and texture must be coded. Coding of both the shape and texture may be intra or predictive.

Just as in earlier systems, the syntax must support many coding possibilities. Any block may be intracoded; in a P-VOP or B-VOP, blocks may be predictively coded. In the simplest case, the *motion vector difference* (MVD) is zero (i.e. the motion-vector prediction is perfect), and the motion vector points to a good match for the block, so that no residuals are required. This block just requires a "skip" code. If the match is not sufficiently good, residuals must be transmitted. In other cases, the MVD will be nonzero and residuals may or may not be required.

Sprites

MPEG-4 has another object type that is particularly useful for backgrounds; the sprite. A *sprite* is a video object, usually larger (perhaps considerably larger) than can be displayed at any one time. It is used for an object (like a static background) that is persistent through a scene. Although sprites are not restricted to games, this application provides their most obvious use. Generally a game scene consists of a background plus a number of (usually) synthetic objects that move according to the script of the game and the actions of the player. During the action, the viewed scene will likely pan around the background; the player sees different parts at different times, but they are all parts of the same static image. MPEG-4 provides the ability to transmit the whole of the background once as a sprite, and to generate scenes by sending cropping or warping information to determine which part of the sprite is visible at any given instant (Figure 11-3). Once the sprite has been transmitted, only the cropping/warping information for the sprite and the foreground objects need be transmitted. In a typical game, each section of the sprite will likely be used many times, so this approach represents a substantial reduction in the total data used.

Figure 11-3
The sprite provides a
background scene
larger than the
display window;
once all the sprite
has been transmitted,
any part of it may
be displayed.

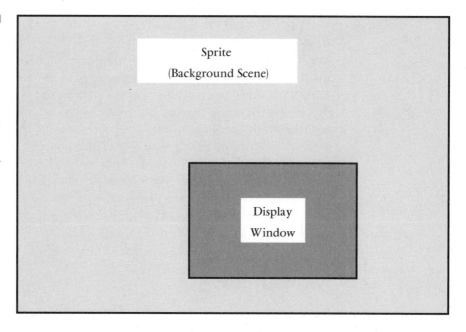

Transmitting the whole of a sprite at the beginning may be very efficient, but will need either extra bandwidth or a period of time before the action can start. MPEG-4 has techniques to avoid this problem as well. A sprite can be transmitted in sections, as needed. The opening scene would require that only the area of the sprite needed for that scene be transmitted immediately. If the "camera" then panned to the left, a new strip of information would be transmitted every frame, but these would all be stored at the decoder as part of the sprite. So, if the pan reversed, no new background information would need to be transmitted until the camera moved to the right of the original scene. However, a certain amount of sprite data could be transmitted with each frame anyway, as bandwidth permitted, to continue building the complete image at the decoder. This method still requires that a full screen of data be transmitted before the first scene may be viewed. If this is unacceptable, the sprite may be *progressively* encoded, so that only a low-resolution version is transmitted at first; more area and more resolution are sent later.

Sprites are encoded as luminance plus two color components, as we have seen in JPEG and earlier versions of MPEG, and are always intra-coded, since the image is essentially static. However, I-encoding uses the improved techniques introduced in MPEG-4, discussed in the previous section.

Static Texture Coding

MPEG-4 includes the ability to map a static texture onto a varying shape. For example, a logo or message may be mapped onto a swimming fish. This operation is performed at the decoder; the moving shape and the static texture are transmitted as separate objects. In this case the ability to scale the static texture is very important. The movement of the shape may cause stretching or expansion of the texture far beyond its nominal size. To maximize the quality of this operation, MPEG-4 provides a wavelet-based tool for compression of the static texture. Wavelets are discussed in a later chapter; the important point here is that wavelets provide good scaling capability; objects that are expanded display minimal artifacting.

Animations

As discussed above, one of the great strengths of MPEG-4 is the ability to transmit both natural and synthetic objects that may be composited at the decoder. One of the more interesting capabilities using synthetic objects is *facial animation*. This is another example of mapping a texture onto a moving shape, but in this case the shape is specified by a mesh or 3-D model specified by a number of *nodes*. The position of each node is coded, and again predictive coding may be used to increase the efficiency of the coding as the face-shape changes. The standard includes all the syntax necessary to code the position and movement of the nodes. It also includes higher-level constructs, such as *visemes*, or visual lip configurations equivalent to speech phonemes.

With sufficiently good implementation, this technique opens up some fascinating (or frightening) possibilities. A real or imaginary person may be represented by a mesh and a texture. In combination with speech synthesis, the animated face can be coded to "read" any textual message.

Version 2 of MPEG-4 goes on to add body animation capability. The following extract from the official Overview of the MPEG-4 Standard is reproduced by kind permission of the MPEG Convener. Not only does this short quote give a description of the body-animation capability, but I think it provides an excellent example of the thoroughness of every aspect of MPEG-4.

The Body is an object capable of producing virtual body models and animations in the form of a set of 3-D polygonal meshes ready for rendering. Two sets of parameters are defined for the body: body definition parameter (BDP) set, and body animation parameter (BAP) set. The BDP set defines the set of parameters to transform the default body to a customized body with its body surface, body dimensions, and (optionally) texture. The body animation parameters (BAPs), if correctly interpreted, will produce reasonably similar high level results in terms of body posture and animation on different body models, without the need to initialize or calibrate the model.

Upon construction, the Body object contains a generic virtual human body with the default posture. This body can already be rendered. It is also immediately capable of receiving the BAPs from the bitstream, which will produce animation of the body. If BDPs are received, they are used to transform the generic body into a particular body determined by the parameters contents. Any component can be null. A null component is replaced by the corresponding default component when the body is rendered. The default posture is defined by standing posture. This posture is defined as follows: the feet should point to the front direction, the two arms should be placed on the side of the body with the palm of the hands facing inward. This posture also implies that all BAPs have default values.

No assumption is made and no limitation is imposed on the range of motion of joints. In other words the human body model should be capable of supporting various applications, from realistic simulation of human motions to network games using simple human-like models. The work on Body Animation includes the assessment of the emerging standard as applied to hand signing for the listening-impaired.

Scalability

MPEG-4 provides both spatial and temporal scalability at the object level. In both cases this technique is used to generate a *base layer*, representing the lowest quality to be supported by the bitstream, and one or more *enhancement layers.* These layers may all be produced in a single coding operation. The scaling can be implemented in two different ways. When there are known bandwidth limitations, versions of the bitstream may be used that include only the base layer, or the base layer plus lower-order enhancement layers. Alternatively, all layers may be sent and the scaling decision left to the decoder. If the display device is low resolution, or if computational resources are insufficient, enhancement layers may be ignored.

Figure 11-4 shows the concept of an encoder implementing spatial scalability; in this case at just two levels. The input VOP is down-converted to a lower resolution, the base layer. This layer is encoded, and then a decoder constructs the base-layer VOP as it will appear at the decoder. This VOP is then up-converted to the same resolution as the input, and a subtraction process yields the differences from the original image, these are separately encoded in an enhancement-layer encoder. Note that each stream of encoded VOPs forms a video object layer. The base-layer VOL uses both intra- and intercoding, but the enhancement layer uses only predictive coding. The base-layer VOPs are used as references, as shown in Figure 11-5.

Figure 11-4

Spatially scalable
encoder for a single
enhancement layer.

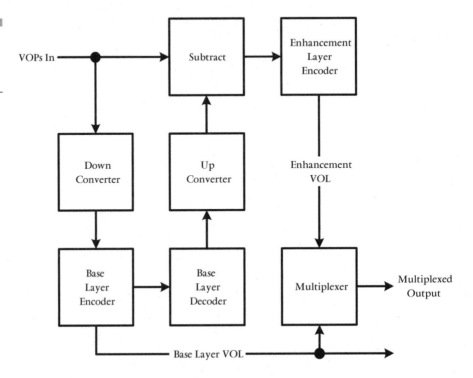

Figure 11-5
Prediction of a spatial
enhancement layer.

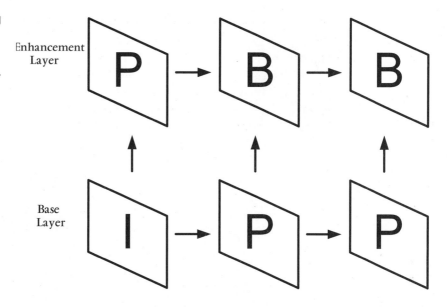

Temporal scalability is rather more simple. The stream of incoming VOPs is split. The required number of VOPs (perhaps every third VOP) is sent to the base-layer encoder; the remainder is sent to one or more enhancement layers.

Note that scalability is applied on a per-object basis. This provides a wide range of coding and decoding capabilities. For example, a decoder in a game system might have insufficient computational power to decode all objects at the highest available rate. It might choose to decode only a low rate for the background, accepting some level of jerky motion there, but to decode a higher rate for the principal foreground object, making its motion as smooth as possible.

Advanced Coding Extensions (ACE)

Version 2 of MPEG-4 introduced three new tools to improve coding efficiency for video objects. These are *global motion compensation* (GMC), *quarter pel* (pixel) *motion compensation*, and *shape-adaptive DCT*. Together, these tools have been shown to improve coding efficiency by up to 50 percent over Version 1, depending on the type of material and the bit rate.

Global motion compensation and quarter pixel motion compensation are fairly self-explanatory. GMC allows the overall motion of the object to be coded with a very few parameters, and the improved resolution of motion vectors substantially reduces prediction errors and the use of residuals.

Shape-adaptive DCT may be used to improve the coding efficiency of boundary blocks, when not all pixels belong to the image. Instead of using 8×8 two-dimensional DCT on the complete block of 64 pixels, one-dimensional DCT is applied first vertically, then horizontally, but only to the pixels that belong to the object; these are called the *active pixels.*

To start the process, each column of the block is examined, and within each column all the pixels that belong to the object are moved up, so the column is "top-justified." Then, each column is DCT-transformed, but using a DCT matched in size to the number of pixels. For example, if there are five pixels in a column that belong to the object, a 1×5 DCT would be used for that column. If there are no active pixels in the column, no transform is performed. When the vertical transforms are complete, the active pixels are shifted to the left, so that each row is "left aligned." A one-dimensional DCT is then performed on each row that has active pixels, again using the appropriate size of transform.

Because the contour of the object is transmitted separately by shape coding, the pixels can be moved back to the "the right place" in the decoder.

Visual Profiles

MPEG-4 includes a structure of profiles and levels using the same philosophy as MPEG-2. As with all aspects of MPEG-4 the range is vast. Rather than paraphrase, I include part of the listing from the Official Overview; this extract covers visual profiles only for Versions 1 and 2. Figure 11-6 shows the relationships among the various profiles and includes the two studio profiles defined by the proposed Amendment 3, and the Fine Grain Scalability profile that is the subject of proposed Amendment 4. These additions to the standard are discussed below. Figure 11-6 is reproduced from the work of Professor Jens-Rainer Ohm of the University of Aachen, with his kind permission.

> The visual part of the standard provides profiles for the coding of natural, synthetic, and synthetic/natural hybrid visual content. There are five profiles for natural video content:

Figure 11.6
Profiles in MPEG-4.

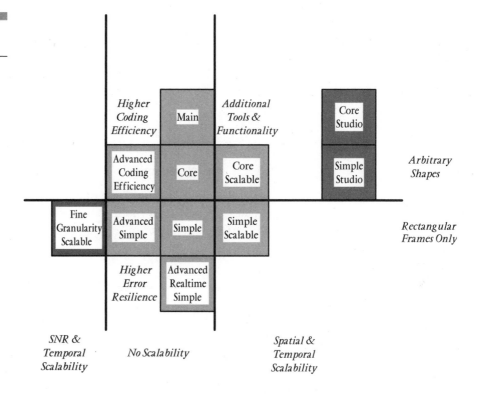

1. The *Simple Visual Profile* provides efficient, error-resilient coding of rectangular video objects, suitable for applications on mobile networks, such as PCS and IMT2000.

2. The *Simple Scalable Visual Profile* adds support for coding of temporal and spatial scalable objects to the Simple Visual Profile. It is useful for applications which provide services at more than one level of quality due to bit-rate or decoder resource limitations, such as Internet use and software decoding.

3. The *Core Visual Profile* adds support for coding of arbitrary-shaped and temporally scalable objects to the Simple Visual Profile. It is useful for applications such as those providing relatively simple content interactivity (Internet multimedia applications).

4. The *Main Visual Profile* adds support for coding of interlaced, semitransparent, and sprite objects to the

Core Visual Profile. It is useful for interactive and entertainment-quality broadcast and DVD applications.

5. The *N-Bit Visual Profile* adds support for coding video objects having pixel depths ranging from 4 to 12 bits to the Core Visual Profile. It is suitable for use in surveillance applications.

The profiles for synthetic and synthetic/natural hybrid visual content are:

1. The *Simple Facial Animation Visual Profile* provides a simple means to animate a face model, suitable for applications such as audio/video presentation for the hearing impaired.

2. The *Scalable Texture Visual Profile* provides spatial scalable coding of still image (texture) objects useful for applications needing multiple scalability levels, such as mapping texture onto objects in games, and high-resolution digital still cameras.

3. The *Basic Animated 2-D Texture Visual Profile* provides spatial scalability, SNR scalability, and mesh-based animation for still image (textures) objects and also simple face object animation.

4. The *Hybrid Visual Profile* combines the ability to decode arbitrary-shaped and temporally scalable natural video objects (as in the Core Visual Profile) with the ability to decode several synthetic and hybrid objects, including simple face and animated still image objects. It is suitable for various content-rich multimedia applications.

Version 2 adds the following Profiles for natural video:

1. The *Advanced Real-Time Simple* (ARTS) Profile provides advanced error resilient coding techniques of rectangular video objects using a back channel and improved temporal resolution stability with the low buffering delay. It is suitable for real time coding applications; such as the videophone, tele-conferencing, and remote observation.

2. The *Core Scalable Profile* adds support for coding of temporal and spatial scalable arbitrarily shaped objects to the Core Profile. The main functionality of this profile is object-based SNR and spatial/temporal scalability for regions or objects of interest. It is useful for applications such as the Internet, mobile, and broadcast.

3.. The *Advanced Coding Efficiency* (ACE) Profile improves the coding efficiency for both rectangular and arbitrary shaped objects. It is suitable for applications such as mobile broadcast reception, the acquisition of image sequences (camcorders), and other applications where high coding efficiency is requested and small footprint is not the prime concern.

The Version 2 profiles for synthetic and synthetic/natural hybrid visual content are:

1. The *Advanced Scalable Texture Profile* supports decoding of arbitrary-shaped texture and still images including scalable shape coding, wavelet tiling, and error-resilience. It is useful for applications the require fast random access as well as multiple scalability levels and arbitrary-shaped coding of still objects. Examples are fast content-based still image browsing on the Internet, multimedia-enabled PDAs, and Internet-ready high-resolution digital still cameras.

2. The *Advanced Core Profile* combines the ability to decode arbitrary-shaped video objects (as in the Core Visual Profile) with the ability to decode arbitrary-shaped scalable still image objects (as in the Advanced Scaleable Texture Profile.) It is suitable for various content-rich multimedia applications such as interactive multimedia streaming over Internet.

3. The *Simple Face and Body Animation Profile* is a superset of the Simple Face Animation Profile, adding—obviously—body animation.

Scene Compositing and Interaction

Scene Modeling

Although it is not directly related to video compression, it would be wrong to discuss MPEG-4 without some mention of *BInary Format for Scenes* (BIFS). The introduction to this chapter highlighted the fact that MPEG-4 can handle multiple video objects, all transmitted independently, but composited at the decoder. The same philosophy is applied to audio. A complete MPEG-4 scene may include multiple audio objects, natural and synthetic, such as background sound, dialogue, sound effects, etc. There can also be multiple 2-D and 3-D graphical objects in a scene.

BIFS is the language that describes, statically and dynamically, how objects should be brought together at the decoder to form a complete scene. It is based on *virtual reality modeling language* (VRML) with extensions that provide simpler constructs for 2-D objects and coordinate spaces. BIFS provides a hierarchical or tree structure where objects may be combined into groups. Individual objects may be manipulated within a group, or the whole group may be manipulated. Nodes of the tree may be added or removed at any time.

Each object and each group have a local coordinate system, addressing both space and time parameters. All of these coordinate systems relate to a set of global coordinates describing the scene, so that any object may be manipulated with respect to itself, or within any layer of a group structure, or with respect to the overall scene. Thus, BIFS provides a timeline within which any part of the scene can be choreographed. BIFS data is sent in a separately identified stream or multiplexed with video or audio sessions.

MPEG-4 Version 2 includes some extensions to BIFS, known as *advanced BIFS*. To demonstrate the richness of MPEG-4, I again quote a paragraph from the Official Overview. To stress that the visual schemes discussed above are just one part of MPEG-4, this extract refers to Advanced BIFS constructs for the audio parts of a scene:

> Advanced sound environment modeling in interactive virtual scenes, where properties such as room reflections, reverberation, Doppler effect, and sound obstruction caused by objects appearing between the source and the listener are computed for sound sources in a dynamic environment in real time. Also enhanced source directivi-

ty modeling is made possible enabling inclusion of realistic sound sources into 3-D scenes.

Interaction

MPEG-4 is designed for use in applications where the end-user may interact with the scene presented. The most obvious application example is games, but interaction is a powerful tool that is an essential element of any system where more information is available than can be presented at one time. Interactive television, educational systems, any sophisticated data retrieval system; all need the ability to change the presentation based upon user requests—just as on a Web site.

Interaction may be provided using many devices—mice, remote controls, and keyboards are just a few. Fortunately, the MPEG-4 constructs do not require additional standardization or a guess at future mechanisms for interaction. Any interactive device is required to produce BIFS commands, and these are combined with the BIFS content of the master scene. The decoder need not know whether the commands were part of the original coding or contributed by an interaction mechanism; the same capabilities exist.

Work in Progress

At the time of writing (September 2000) the MPEG committee is completing work on two more amendments to MPEG-4, scheduled to become part of the standard during 2001. As with earlier work, the new amendments are true supersets, extending the application of the Standard into new areas.

Studio Profiles

I mentioned at the beginning of this chapter that MPEG-4 had expanded far beyond its original focus of low-bit-rate applications. *Studio profiles* are intended for use in production environments and for the storage and transport of program material up to the point of transmission to the end-user, where an appropriate profile and level would be chosen for the application. It is important to note that the

MPEG-4 studio profiles are true members of the MPEG-4 system, not just extensions of MPEG-2. In fact, the object-oriented structure is seen as a natural fit with the production process. If, for example, video objects can be kept separate, and their compositing is described separately, there are no irrevocable steps. The original material may be re-edited or repurposed without limitation.

The MPEG-4 studio profiles overlap MPEG-2, in that they include the capabilities of the MPEG-2 4:2:2 profile. MPEG-4, however, goes much further, and the studio profiles include most of the powerful tools and constructs specified for the other profiles, together with extension into higher-definition formats and other areas critical to high-quality production.

The studio profiles add a few more tools to improve efficiency in the high-quality domain. In particular, DCT coefficients may be grouped by value, and variable-length coding uses recursive selection of the most efficient of a number of VLC coding tables. These tools offer an increase in coding efficiency of around ten percent when compared to MPEG-2.

Two studio profiles are specified. Both provide for 4:2:2 or 4:4:4 coding in either the *YUV* or *RGB* domains. They provide for picture sizes of up to 4k×2k (4000×2000 pixels), progressive or interlace, and pixel depths of up to 12 bits per component. Encoded data rates can be up to 1.2 Gbits/s. Both studio profiles support arbitrarily-shaped video objects.

The *Simple Studio Profile* is intended for acquisition and editing environments. The profile permits only intracoding, so that frames or VOPs are always independent, and the video can be accessed at any point by an editor without complication or loss of temporal granularity. The Simple Studio Profile also permits lossless transcoding from the 4:2:2 profile of MPEG-2.

The *Core Studio Profile* is intended for distribution and storage within production environments. It includes all of the elements of the Simple Studio Profile, plus sprites, and permits predictive encoding, using P-VOPs only. There is no bidirectional prediction within MPEG-4 studio profiles.

Fine Grain Scalability

The *Fine Grain Scalability Profile* (FGS), often known as the *Streaming Profile* is intended to support applications and environments where the bandwidth and/or computational power available cannot be predicted and may vary dynamically. The scaling tools provided by other parts of MPEG-4 are powerful in their own way, but have two significant limitations.

The granularity of scaling using existing tools is necessarily coarse. Earlier, we saw a process that could produce a 2-layer bitstream including a base layer and one enhancement layer. The number of enhancement layers that can be used is arbitrary, but in practical terms is severely limited. Each layer adds coder and decoder complexity and overhead in the bitstream. Also, provision for all the levels of scalability must be built in at the time of encoding; any changes require recoding from scratch.

In contrast, FGS provides a mechanism that permits a single encoding process, producing a bitstream that can be modified subsequently in two different ways. Prior to transmission, the bitstream may be processed to scale it down to known bandwidth limitations. This can be performed dynamically, for example in response to the requirements of a statistical multiplexing system. Downstream distribution points may further reduce the bit rate if necessary. At the receiving end the decoder can use all of the received bits, or it can scale the bitstream down further to match its capabilities and resources, again dynamically.

Figure 11-7 shows the concept of bit-plane encoding. The quantized DCT coefficients are indicated by the columns of bits. Clearly the most significant coefficient is represented by the tallest column. We can "slice" the bit planes horizontally. If we took only the highest plane in the figure, we would capture only the MSB of the largest coefficient. Our representation of the block would then be a single coefficient with the quantized value 1000 (instead of 1110). This is obviously a very crude representation, but it is better than nothing.[2] If we were able to take two bit planes, we would get three coefficients, one with the value 1100, and two each with the value 100. The important point is that from a single coding operation we can extract at the decoder as much information as we are able to use from the bitstream, and be assured that we have captured the most significant elements. This approach is also known as *signal-to-noise scalability*.

It is possible to modify this technique at the encoder. Some values may be artificially pushed up the bit-plane stack. For example, it is possible to apply frequency weighting to the DCT coefficients,

---

[2] As discussed in Chapter 6, the reconstruction value would likely be different. For an effective range of 1000 to 1111, the probable reconstruciton value is 1100.

emphasizing the importance of the low-frequency coefficients, or artificially to increase the value of all coefficients that lie within a particular *region of interest* in the object.

The streaming profile is, of course, especially useful in Internet environments. Fine-grain scalability may be used alone, or combined with temporal scaling. A session may be multicast by a server at the highest quality permitted by the bandwidth available. Downstream servers may, if necessary, trim the bitstream to suit lower-bandwidth parts of the network. A decoder uses only as much of the bitstream as its resources permit. It may ignore temporal enhancement layers to minimize the number of VOPs to be processed, and it can take only as much data as it can handle from the other layers.

Figure 11-7
Bit planes in an array of DCT coefficients.

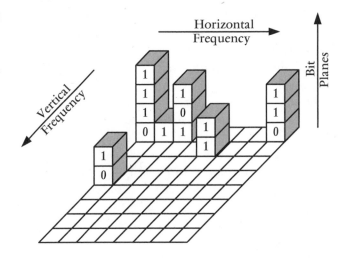

Future Work

A technology area of enormous interest today is *digital cinema*. Although film will likely remain the capture medium of choice for moviemakers for many years, it is increasingly clear that the industry will move toward digital processing, delivery, and projection of movies. Many of the MPEG-4 constructs, particularly those in the studio profiles, work well for digital cinema, but it is most unlikely that all of the requirements will be met by work done to date. In October 2000 the MPEG Committee is expected to publish a formal Call for Requirements for digital cinema, and the responses to this will form the basis for a new work statement to expand MPEG-4 yet further.

Conclusion

I think this is the longest chapter in the book, and those who are familiar with MPEG-4 will know that I have only scratched the surface of this Standard and its various amendments. I have tried to cover, at least briefly, those elements of the standard most relevant to video compression, and to convey an impression of the vast scope of MPEG-4. For those interested in learning more, MPEG has become much more open to accessibility of documents (prior to the final Standard, which is controlled by ISO), and its Web site carries many working documents and tutorials.

CHAPTER 12

MPEG-7
and MPEG-21

Introduction

In one sense, neither MPEG-7 nor MPEG-21 belongs in this book, because neither is about compression. However, the work of the MPEG committees is important as a continuum, and it is useful to understand how its various projects are related. It is already evident that MPEG-7 will become a critically important part of the digital media scene. MPEG-21 is less well defined at the time of writing, but the track record of the committee is excellent, so it will likely evolve into an essential part of the world's future digital infrastructure.

MPEG-7

The obvious first question about MPEG-7 is "why seven?" As we have seen, the cancellation of MPEG-3 caused the sequence of real standards to be MPEG-1, MPEG-2, MPEG-4. There were those on the committee who took the pragmatic approach and expected the next standard to be MPEG-5. The "binary buffs" saw the historical 1—2—4 as the start of a preordained binary sequence, and wanted the new work to be MPEG-8. Finally, it was concluded that any simple sequence would fail to signal the fundamental difference between the new standard and the work of MPEG-1 through MPEG-4; thus MPEG-7 was chosen.[1] It is stated that no one advocated the use of MPEG-6!

Concepts of MPEG-7

Having covered the issue of numbering, we can now address the real issue; what is MPEG-7 about? MPEG-7 is about *metadata*, also known as "bits about the bits." Metadata is digital information that describes the content of other digital data. In modern parlance, the program material or content, actual image, video, audio, or data objects that convey the information, are known as *data essence*. The metadata tells the world all it needs to know about what is in the essence.

---

[1] At least the mathematicians can claim some victory; the sequence 1, 2, 4, 7 is a second-order arithmetic sequence. But the next member is eleven—so much for that theory!

Anyone who has been involved with the storage of information, be it videotapes, books, music, or whatever, knows the importance and the difficulty of accurate cataloging and indexing. Stored information is useful only if its existence is known, and if it can be found when needed. More stringently, it is only really useful if it can be found and retrieved in a length of time that matches that of the application. If a journalist is composing a news story, the ability to refer to or include parts of historical video footage may be very helpful, but not if the story is to air today, and the historical footage will be available in three weeks. It has been said of the television news environment that once the individual journalist who created a tape has forgotten what is on it, the chance of the record being useful again is minimal. Quite simply, it is unlikely that it will be found in time to be useful.

This problem has always been with us and is addressed in the analog domain by a combination of labels, catalogs, card indexes, and other devices. More recently, the computer industry has given us efficient, cost-effective, relational databases that permit powerful search engines to access stored information in remarkable ways, provided, that is, the information exists, and is in a form that the search engine can use effectively.

Here is the real problem: The world is generating new media content at an enormous and ever-increasing rate. With the increasing quantity and decreasing cost of digital storage media, more and more of this content can be stored. Local and wide area networks can make the content accessible and deliverable if it can be found. The search engines can find what we want, and the databases can be linked to the material itself, so that the solution is apparent if we can get all the necessary indexing information into the database in a form suitable for the search engine.

How can we generate that information in a suitable form? Let's look at just some of the data we might want to generate about a piece of news footage:

- The date and time the footage was created
- The location; perhaps the direction the camera was pointing, perhaps even the zoom angle (wide shot or closeup)
- Who or what is pictured; what is happening
- Who owns the footage; who has rights to access it
- Original footage or edited? If edited, references to stored originals

- Video format, soundtrack, commentary
- Storage format
- A unique identifier for the material

These are just a few of the items that may be of interest. The recently published SMPTE Metadata Dictionary defines many hundreds of *data elements* that may be given values to help classify content.

There are two main issues in the generation of metadata. The most obvious is its sheer volume; if indexing must be performed by a human interpreter, resources exist to index only a minute portion of the content being created, without any possibility of indexing historical records. So, some form of automated analysis of digital records is essential. This presents enormous technical challenges. Solutions will arrive only gradually, but we can be sure they will arrive. The other element is standardization of descriptions. Is a location to be specified by zip code, by GPS coefficient data, or by a hierarchical country—state—city—street construct, or by a contextual reference, such as White House Rose Garden?

We might guess from knowledge of earlier standards that the MPEG committee would not concern itself unduly with mechanisms for generating data. MPEG rightly takes the view that if it creates a standardized structure, and if there is a market need, the technological gaps will be filled. In previous MPEG standards, the syntax and the decoder were specified by the standard; the generation of the syntactical elements (the operation of the encoder) was left to the implementer. In MPEG-7, *only* the syntax is standardized. The generation of the metadata is unspecified, as are the applications that may use the metadata. For any device that generates metadata, MPEG-7 specifies how that information should be expressed. Thus, the fields that should go into a database are specified, and anyone designing a search engine knows what descriptive elements may be present and how they will be encoded.

Figure 12-1 shows the scope of MPEG-7. Even with this limited scope, the goals of MPEG-7 are wide ranging. The following is a list, taken from the MPEG-7 FAQ, of example queries that a search engine accessing an MPEG-7 database should be able to process:

1. **Music:** Play a few notes on a keyboard and get in return a list of musical pieces containing (or close to) the required tune or images somehow matching the notes, e.g. in terms of emotions.

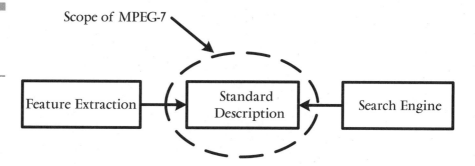

2. **Graphics:** Draw a few lines on a screen and get in return a set of images containing similar graphics, logos, ideograms, etc.

3. **Image:** Define objects, including color patches or textures and get in return examples from among which you select interesting objects to compose your image.

4. **Movement:** On a given set of objects, describe movements and relations between objects and get in return a list of animations fulfilling the described temporal and spatial relations.

5. **Scenario:** On a given content, describe actions and get a list of scenarios where similar actions happen.

6. **Voice:** Using an excerpt of Pavarotti's voice, and getting a list of Pavarotti's records, video clips where Pavarotti is singing or video clips where Pavarotti is present.

MPEG-7 Terminology

The following short extract from the Official Overview defines the terminology used in the standard:

- **Data:** Data is audio-visual information that will be described using MPEG-7, regardless of storage, coding, display, transmission, medium, or technology.

- **Feature:** A feature is a distinctive characteristic of the data which signifies something to somebody.

- **Descriptor:** A descriptor (D) is a representation of a feature. A descriptor defines the syntax and the semantics of the feature representation.

- **Descriptor Value:** A descriptor value is an instantiation of a descriptor for a given data set (or subset thereof).

- **Description Scheme:** A description scheme (DS) specifies the structure and semantics of the relationships between its components, which may be both descriptors and description schemes.

- **Description:** A description consists of a DS (structure) and the set of descriptor values (instantiations) that describe the data.

- **Coded Description:** A coded description is a description that has been encoded to fulfill relevant requirements such as compression efficiency, error resilience, random access, etc.

- **Description Definition Language:** The Description Definition Language (DDL) is a language that allows the creation of new description schemes and, possibly, descriptors. It also allows the extension and modification of existing description schemes.

MPEG-7 Structure

From the definitions above, we can see how MPEG-7 structures work. A *descriptor* (D) is the lowest level of the hierarchy and specifies how the value of a feature is expressed. It usually represents some relatively simple concept; anything complex should be broken down into its component parts, each of which should have a descriptor. Even though the concept should be simple, descriptors may be elementary or sophisticated. A descriptor might refer to a primitive, such as a circle shape, or to a much higher-level concept such as a person's name. A *description scheme* (DS) "collects" descriptors into useful sets that represent some useful higher-level entity. Description schemes may themselves contain other DSs.

Let's look at a possible example, but it must be stressed that the example is hypothetical to illustrate the concept—it probably is not a valid MPEG-7 construct, and the technology may not exist to implement it.

Descriptors may be defined for shape, color, texture, and other attributes. One DS might specify a set of descriptor values, or ranges of values, that together represent a human face. Other DSs could be defined that represent other elements of the human body. Yet another DS could include all of those DSs in specified spatial relationships to construct a generic human body. If we had a way of specifying the values appropriately, a *description* could use this DS with a particular set of values to represent one specific person. Using a query-by-example approach for a search (not part of the standard), and a file description of a person, it should be possible to search a database to identify video records that show that person.

Description schemes and descriptors are written using the *Description Definition Language* (DDL). This is a language based on *eXtensible Markup Language* (XML), with extensions defined specifically for MPEG-7.

It should be noted that descriptions can be applied to "conventional" video and audio, or to video and audio objects, as used in MPEG-4 constructs. Any separation of the video and audio into objects is likely to make MPEG-7 descriptions easier to achieve and more useful.

MPEG-7 Visual

The MPEG-7 Visual document defines four areas for description: color, texture, shape, and motion. Within each category there are elementary and sophisticated descriptors. The descriptors are based on two structures; the *grid layout* and the *histogram*, which can be used separately or in combination. The grid layout permits definition and analysis of any rectangular segment or region of the image or image sequence. For example, a region could be the top-left corner of the image, over a sequence of frames or VOPs. The histogram can provide analysis of any measurable characteristic over the whole image, or a specific region.

Color descriptors include:

- *Color space*, such as *YUV* or *RGV*[2]—the scheme by which color is described. This is usually used in conjunction with one of the color descriptors.

---

[2] Most input data will be in *RGB* or *YUV* form, but these may be transformed into other color spaces for purposes of describing the image. To this end, MPEG-7 also supports HSV (Hue, Saturation, Value) and HMMD (Hue, Min, Max, Difference) that may be computed from *RGB* or *YUV*. These latter color spaces more closely emulate the human visual model and are expected to provide better search data.

- *Dominant color(s)*, to represent objects or regions where a small number of colors (derived by quantization) can accurately characterize the information.

- The *color histogram*, a flexible descriptor that is used in conjunction with color quantization (typically to 64 colors) to characterize an image or region.

- *Color quantization*, used in conjunction with other tools to provide a usefully small number of colors to describe the image. Very sophisticated, non-linear, quantizers may be used.

The list goes on to include more complex tools designed to extract descriptions most appropriate for image searching and matching.

Texture has been found to be a critical element in image recognition. MPEG-7 includes descriptors for edges and for homogeneous textures. Homogeneous textures can be described by two mechanisms. A *texture browsing* descriptor uses complex filtering processes to characterize the regularity and coarseness of textures in one or two dominant directions. This descriptor uses only a small number of bits. The *homogeneous texture* descriptor uses 30 filters and complex subband analysis to produce a 62-byte representation of a texture for similarity searching.

Shape descriptors include *boundary box, region-*, and *contour-based* tools to describe the shape of objects, even if they are not consistently represented in the image. For example, the tools can still provide valuable recognition data even if the apparent shape changes because of movement or occlusion.

Finally, motion tools can describe six-axis *camera motion* plus *zoom, object motion trajectory, parametric motion* of objects, and *motion activity*—used to characterize, for example, the scoring events within a sports sequence.

Summary

Obviously, this is only the most minimal summary of just the visual tools within this ongoing work. MPEG-7 includes a range of tools for audio (including speech recognition and descriptors to characterize musical instruments), and a multimedia descriptive framework that is described by the following extract from the Official Overview:

- *Creation and Production Meta information* describing the creation and production of the content: typical features include title, creator, classification, purpose of the creation, etc. This information is author generated most of the time because it cannot be extracted from the content.

- *Usage Meta information* related to the usage of the content: typical features involve rights holders, access right, publication, and financial information. This information may very likely be subject to change during the lifetime of the AV content.

- *Media Description* of the storage media: typical features include the storage format, the encoding of the AV content, elements for the identification of the media. Note that several instances of storage media for the same AV content can be described.

- *Structural aspects:* Description of the AV content from the viewpoint of its structure: the description is structured around segments that represent physical spatial, temporal, or spatiotemporal components of the AV content. Each segment may be described by signal-based features (color, texture, shape, motion, audio features) and some elementary semantic information.

- *Conceptual aspects:* Description of the AV content from the viewpoint of its conceptual notions. (Note that currently this part of the [Multimedia Description Scheme] is still under Core Experiment and no elements are included in the [Experimental Model] or [Working Draft]).

MPEG-7 is an enormously ambitious project, and the work is still in process; in fact there is still considerable contention around many aspects of the proposed standard. However, work is expected to proceed to an International Standard before the end of 2001. The documents and tutorials on the MPEG Web site provide much more information on MPEG-7 for those interested.

MPEG-21

MPEG-21 also differs in concept from the earlier work of the committee. The numbering is said to refer to its anticipated use in the twenty-first century.

Although there is still some considerable confusion about what exactly MPEG-21 is, the basic concept is fairly simple—although wide reaching. MPEG-21 seeks to create a complete structure for the management and use of digital assets, including all the infrastructure support for the commercial transactions and rights management that must accompany this structure. The vision statement is "to enable transparent and augmented use of multimedia resources across a wide range of networks and devices." MPEG-21 starts by defining two fundamental concepts.

A *digital item* is the fundamental unit of distribution and transaction within the MPEG-21 framework. It is a structured digital object represented and identified in a standard manner. In other words, it includes some data essence and the metadata required to describe it and to provide the information necessary to regulate its use.

An MPEG-21 *user* is defined as any entity that interacts in the MPEG-21 environment or makes use of a digital item. Users are individuals and organizations, including creators, consumers, rights holders, content providers, distributors, and others. Note that in MPEG-21 terminology the term "user" does not necessarily mean a consumer; content creators and distributors are also "users."

The Draft Technical Report, published in September 2000, represents the first step in the formal MPEG-21 process. It describes the problem and references a number of possible user-scenarios. These range from editing, distributing, and commercial printing of home snapshots, to complex structures of medical records with mechanisms for access and security. The Report tentatively identifies seven architectural elements expected to form parts of the MPEG-21 multimedia framework:

- The *digital item declaration* is expected to "establish a uniform and flexible abstraction and interoperable schema for defining digital items." The schema must be open and extensible to any and all media resource types and description schemes, and must support a hierarchical structure that is easy to search and navigate.

- The *digital item representation* of MPEG-21 is the technology that will be used to code the content and to provide all the mechanisms needed to synchronize all the elements of the content. It is expected that this layer will reference MPEG-4.

- *Digital item identification and description* will provide the framework for the identification and description of digital items (linking all content elements). This will likely include the description schemes from MPEG-7, but must also include "[a] new generation of identification systems to support effective, accurate, and automated event management and reporting (license transactions, usage rules, monitoring, and tracking, etc.)". It must satisfy the needs of all classes of MPEG-21 users.

- *Content management and usage* must define interfaces and protocols for storage and management of MPEG-21 digital items and descriptions. It must support archiving and cataloging of content while preserving usage rights, and support the ability to track changes to items and descriptions. This element of MPEG-21 will likely also support a form of "trading," where consumers can exchange personal information for the right to access content, and formalization of mechanisms for "personal channels" and similar constructs.

- *Intellectual property management and protection* is an essential component. The current controversies surrounding MP3 audio files demonstrate the need for new copyright mechanisms cognizant of the digital world. It can be argued that content has no value unless it is protected. MPEG-21 will build on the ongoing work in MPEG-4 and MPEG-7, but will need extensions to accommodate new types of digital items, and new delivery mechanisms.

- MPEG-21 *terminals and networks* will address delivery of items over a wide range of networks, and the ability to render the content on a wide range of terminals. Conceptually, a movie should be deliverable in full digital-cinema quality to a movie theater, or at a lower quality over a slower network to a consumer device (at a different price). In either case there will be some restriction on the type and or number of uses. The user should not be aware of any issues or complexities associated with delivery or rendering.

■ Finally, there is a need for *event reporting* to "standardize metrics and interfaces for performance of all reportable events." The most obvious example here is that if the system allows a user access to a protected item, it must also ensure that the appropriate payment is made.

The Draft Technical Report and the Use Case Scenarios for MPEG-21 were published in September 2000, and a Call for Proposals was issued by the October 2000 meeting of the MPEG committee. MPEG-21 is expected to reach the Committee Draft stage by December 2001.

Pro-MPEG and MPEG Operating Ranges

Introduction

Over the course of the last few chapters we have examined the fundamentals of MPEG and seen its evolution from MPEG-1 through to the present day. There is no doubt that MPEG is the system of choice for delivery of compressed video, and this is likely to continue for many years. Delivery, usually to a home consumer, is exactly what MPEG was designed to do. It was conceived as a one-time process. Video is compressed and delivered to the consumer, where it is decompressed and viewed.

Things have also been changing in the television studio environment, however. The advent of high-quality compression has made possible cost-effective disc and tape recorders that record a compressed data stream and provide excellent video quality. People who a few years ago were adamant that compression would never be used in their plants are now eagerly buying these machines.

Many early disc recorders were based on Motion-JPEG compression, but now all mainstream devices use either MPEG or DV, a different compression system discussed in the next chapter. The first MPEG implementations used an I-B GOP structure at a compressed bit rate of 18 Mbits/s. These machines had limitations for multigeneration studio operations, and this created an opportunity for 50 Mbits/s DV machines to gain a foothold in the marketplace.

This exposed a weakness of MPEG. Because MPEG is designed as a delivery system, in the design phase not much attention was paid to studio applications. This weakness, in fact, stems from MPEG's enormous flexibility: Within the limits of the chosen profile and level, MPEG never specifies exactly how the compression should be performed. Different encoders can and do produce different bitstreams from the same video. More to the point, a signal that has been encoded and decoded may then be presented to a second encoder that applies the MPEG tools differently and makes different assumptions.

Let's look at an example. With long-GOP coding, the first encoder generates I-, P-, and B-frames and makes a set of decisions about how many bits should be allocated to each frame. When this bitstream is decoded, the output is video. The second encoder has no way of knowing which video frames were I-frames in the first coding pass, and which were P-frames or B-frames. So, it is quite likely that the second encoder will form an I-frame out of something that was a B-frame in the previous pass.

Provided the quality of all frames is sufficiently high, this does not present a problem. However, if we are dealing with a transmission chain, where a signal has been contributed over a low-bandwidth link, the B-frames may have been compressed to the extent that there are too many errors to construct an efficient I-frame.

This is just one example of how the flexibility of MPEG can lead to difficulty in attempts to concatenate several MPEG processes, particularly when equipment from several different manufacturers is involved. Remember, MPEG specifies only the syntax and the decoder. It does not specify how the encoding should be performed.

Another issue in studio operations revolves around the use of buffers in MPEG. The existence of transmit and receive buffers makes it possible for different frames to be represented by different numbers of bits, while transmitting over a fixed bit rate channel. Even in the simple case of I-frame-only encoding, with a known nominal bit rate, different encoders may allocate a different number of bits to each frame. To an MPEG proponent, this is an advantage. More bits may be allocated to complex frames, less to simpler frames, thus leading to more consistent quality.

Difficulties arise, however, when recording such a bitstream on tape. A continuous-stream tape recorder must run the tape at a constant speed. It follows that, for any practical design, the head-to-tape speed is also constant, and this implies that the machine records a constant number of bits per second. If video frames are represented by differing numbers of bits, more or less tape will be used for each frame, or each frame must be padded to the same number of bits. In fact, the requirement to edit videotape virtually demands that a constant number of bits be recorded on tape for each frame.

In contrast to MPEG, DV offered some persuasive advantages for the studio environment. We discuss DV in detail in the next chapter, but two points in particular are relevant to this discussion. DV uses I-frame-only encoding, with a constant number of bits per frame. There are no interoperability issues; every DV encoder is the same, and given the same video input, any two DV encoders produce identical encoded bitstreams.

Goals of Pro-MPEG

Manufacturers of MPEG equipment became aware of these issues and saw the need to improve the interoperability of MPEG equipment, particularly in studio environments. This led to the formation of the Professional MPEG Forum, known as Pro-MPEG, in 1998. Two years later, the forum has some 100 members, approximately equally divided between manufacturer and user organizations. The mission of Pro-MPEG includes four principal objectives:

- Encourage interoperability in professional television environments.
- Provide a forum for manufacturers and users to test interoperability.
- Propose guidelines and "Codes of Practice" to accelerate implementation.
- Establish system architectures to maximize interoperability.

The organization meets about four times a year and has staged demonstrations at five industry trade shows in Europe and North America during 1999 and 2000. The Pro-MPEG Forum has established guidelines for *operating ranges* within the MPEG-2 4:2:2 profile, and these have been submitted to the Society of Motion Picture and Television Engineers (SMPTE) for formal documentation as a Recommended Practice. The goal of these operating ranges is to meet the user requirement "to guarantee interoperability through more tightly defined MPEG coding parameters." The following descriptions and figures are reproduced by permission from the proposed SMPTE document. The ranges are shown diagrammatically in Figure 13-1.

Note that at the time of writing the SMPTE document is not finally approved, and changes may take place before publication. Also, the text below presents just an outline of the requirements without necessary additional detail. It is essential that anyone wishing to make use of these operating ranges consult the final SMPTE document when it is published, probably at the end of 2000.

OPERATING RANGE 1: (SDTV, ANY GOP)
Operating range 1 covers SDTV coded at up to 50 Mbits/s and may use temporal predictive coding.

OPERATING RANGE 2: (SDTV I-ONLY)
Operating range 2 covers SDTV coded at up to 50 Mbits/s using no temporal predictive coding. For this operating range, an encoder rate control should ensure that no frame exceeds a limit of 50 Mbits divided by the number of frames per second. For example, at 29.97 I-frames per second, no frame may have more than 1,668,328 bits net data, or at 25 I-frames per second, no frame may have more than 2,000,000 bits net data.

OPERATING RANGE 3A AND 3B (HDTV, ANY GOP)
Operating range 3A covers HDTV coded at up to 80 Mbit/s and may use temporal predictive coding.

Figure 13-1
Proposed operating
ranges for the MPEG
4:2:2 profile.

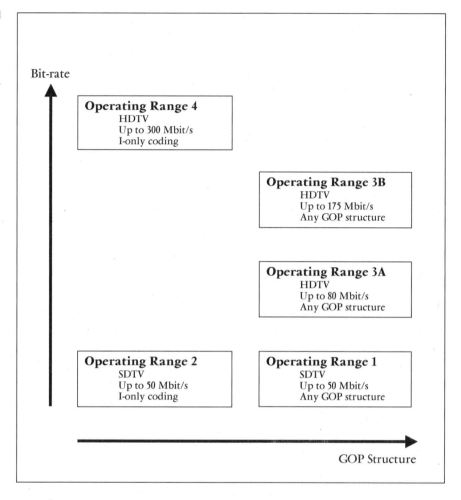

Operating range 3B covers HDTV coded at up to 175 Mbit/s and may use temporal predictive coding.

OPERATING RANGE 4 (HDTV, I-ONLY)

Operating range 4 covers HDTV coded at up to 300 Mbit/s, using no temporal predictive coding. For this operating range, an encoder rate control should ensure that no frame exceeds a limit of 300 Mbits divided by the number of frames per second. For example, at 29.97 I-frames per second, no frame may have more than 10,010,000 bits net data.

Relationships between different Operating ranges are illustrated in Figure 13-2. Operating range 2 is a subset of operating ranges 1 and 4;

Figure 13.2
Relationship between
proposed operating
ranges.

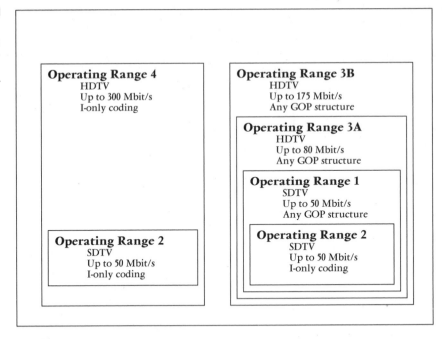

operating range 1 is a subset of operating ranges 3A and 3B, and operating range 3A is a subset of operating range 3B.

Pro-MPEG also attempted to define an *operating point* at 50 Mbits/s I-frame only, to constrain the MPEG parameters even more closely. However, this document became very close to the SMPTE document specifying compression for the D-10 videotape format, an MPEG-based, 50 Mbits/s I-frame only, machine. It was decided that two documents so similar would create confusion, so the operating point document was abandoned.

It is, perhaps, unfortunate that the requirements of a videotape recorder were permitted to constrain a much wider field of operations. Videotape machines have some very special requirements. They are expected to produce pictures while in shuttle mode when the tape is moving at high speed and the movement of the heads does not follow the helical tracks. In this mode, only occasional blocks of data are recovered intact. For this to be useful, dependency between blocks must the reduced as much as possible because the reference block may not be available. To accommodate this requirement, the D-10 document specifies that every slice may contain only one macroblock. This eliminates predictive coding of DC coefficients between macroblocks.

The proposed operating ranges represent the most tangible results of the Pro-MPEG Forum to date, and the work most relevant to this book. However, it is not the only work undertaken by the Forum. Pro-MPEG has also addressed transmission of MPEG streams over Asynchronous Transfer Mode (ATM).

The Pro-MPEG Forum is also working on a universal file format for the exchange of multimedia content. This format, known as *Media eXchange Format* (MXF), is intended to be compression independent and include identification and synchronization information.

DV Compression

Introduction

The DV compression scheme is the product of cooperation among a large number of manufacturers, but with four companies performing most of the development and owning most of the intellectual property. It began, I am told, as an exercise to produce a compression scheme for use in high-definition consumer camcorders, using a data rate of 50 Mbits/s. It soon became apparent that the work presented an opportunity for a new generation of standard-definition camcorders working at 25 Mbits/s and offering the "digital" cachet. The 25 Mbits/s rate was very suitable, because recording at this rate can be implemented quite easily with robust, inexpensive technology.

Even then, DV exceeded its expectations; it is sometimes described as the technology that went from engineers to users before the marketing departments realized what was happening! Not only is it successful in the consumer market, but the technology and its derivatives have been adopted in the profession television broadcast market—in fact, video recorders based on DV it is claimed that have been adopted more rapidly than any other format. Two versions of DV are in use professionally. 25 Mbits/s systems, very similar to the consumer DV products, are used for acquisition, particularly in news environments. A 50 Mbits/s version is used in studios and for postproduction.

Although DV is based on the DCT transform, many aspects are quite different from the MPEG approach. Before looking at the details, it is important to review the design criteria, remembering that the original intention was a consumer product. We have seen that motion estimation is a very expensive process most suitable for an asymmetric environment (few encoders, many decoders). For a consumer product, it was determined that there was no sufficiently economical system of temporal compression, so DV had to be an intraframe system. This approach allows simple editing, also a substantial advantage.

When one is recording on a videotape, it is most helpful if the bit stream to be recorded has a constant number of bits per frame. This was also a design criterion for DV.

MPEG can of course work with I-frames only. However, the MPEG model controls bit rate by a feedback system. It is possible to approximate constant-bit rate compression with MPEG (as specified for the D-10 format mentioned in the previous chapter), but stuffing is generally necessary to achieve an exact fixed bit rate.

MPEG enthusiasts would argue that fixed bit rate is, in fact, undesirable. As we discussed earlier, the human psychovisual system is more sensitive to changes in perceived quality than it is to absolute quality. Constant bit rate with images of varying complexity necessarily implies variable quality of the compressed image. DV proponents, on the other hand, show research results that supposedly demonstrate that DV techniques give a lower minimum quality in a sequence of frames with varying complexity. The arguments will continue!

Finally, the fact that DV was intended for consumer equipment meant cost effectiveness was a prime design goal. It was required that compression be performed on a single chip, and the hope was that the same chip could be used for both compression and decompression. These goals were achieved.

Basic Concepts of DV Compression

DV uses the DCT transform; it also performs quantization of the coefficients, as in JPEG and MPEG. However, DV is specifically designed to generate a fixed number of bits from each frame. It uses some very sophisticated techniques to ensure that bit allocation is optimal for every frame.

The process is described in detail below, but we start with a brief summary. The first step, if necessary, is to reduce the number of color difference samples. For 25 Mbits/s consumer products there are different approaches for 525/60 and 625/50 video. For 525/60 systems the video (normally 4:2:2) is first reduced to 4:1:1. This is an appropriate choice when the video will eventually be coded to NTSC. For 625/50 systems it was decided that DV video should be compatible with the MPEG-2 video transmitted using the Digital Video Broadcasting (DVB) system—adopted in Europe for satellite, cable, and terrestrial transmission—so 4:2:0 coding was selected. For the professional version of DV compression described below, 4:1:1 is used for both 525/60 and 625/50 at 25 Mbits/s; 4:2:2 is used for both standards for 50 Mbits/s compression.

The next step is to break the image up into smaller pieces, each of which is allocated a fixed number of bytes. DV uses macroblocks, much as JPEG and MPEG do. Five macroblocks make a *segment*, and each segment must be compressed to the same number of bytes. The five macroblocks in one segment are taken from different parts of the frame, in an attempt to ensure that no segment gets more than its share of complexity.

Each 8×8 block in the segment is DCT transformed and analyzed for complexity and, depending on the complexity, is allocated to one of four banks of quantization tables. Each bank is optimized for compression of blocks in a particular energy range. There are effectively 16 tables in each bank, and each of these corresponds to a level of quantization scaling. The system then quantizes the coefficients of all blocks in the segment, using a table from the correct bank for each block. This is performed for each of the 16 possible quantization-scaling factors, and the system selects the scaling factor closest to, but not exceeding, 385 bytes of quantized coefficients. Thus, each segment is compressed to an equal number of bytes, but within that segment each block is quantized using a table appropriate to the energy of that block. Complex blocks receive a greater share, and simple blocks a lesser share, of the bits available. There is a third mechanism that helps distribute the bits optimally; this is discussed in the more detailed description that follows.

Detailed Description

25 Mbits/s Compression

DV compression for consumer products is specified by IEC document 61834. The DV-based compression used for professional products described in this section is specified by SMPTE Standard 314M, and figures used in this section are reproduced from this standard, by kind permission of the Vice President, Engineering, of SMPTE. These two documents are the definitive descriptions of DV compression and recommended reading for anyone working with DV video.

Prior to compression to the 25 Mbits/s professional standard, the signal must be reduced from the (normal) 4:2:2 coding to 4:1:1 coding. The DV standard does not specify any filtering, but merely discards pixels from the 4:2:2 input, as shown in Figure 14-1 and Figure 14-2. Depending on the source of the images, appropriate prefiltering may be necessary to prevent artifacts being generated by this decimation process.

■■■ ■■■ ■■

Figure 14-1

Reduction to 4:1:1
sampling for 525/60
systems.

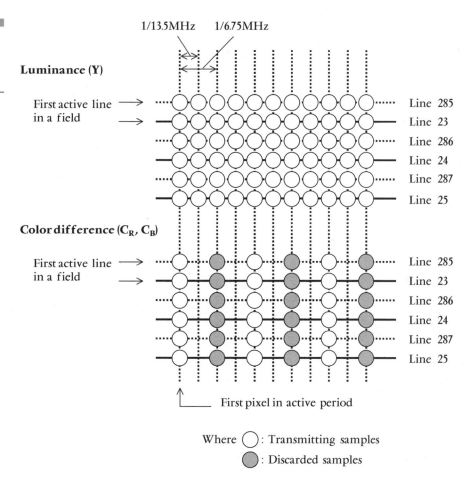

The image is then divided into blocks and macroblocks. In the consumer 625/50 standard with 4:2:0 coding, this process is straightforward, and the same as MPEG. 4:1:1 coding, however, presents a slight problem, because there are only 180 samples of each color difference signal on a line, and 180 is not divisible by 8; we are left with a "half block" at the end of the lines. This is resolved by creating a special "end" macroblock, as shown in Figure 14-3.

The macroblocks are grouped into *superblocks*, as shown in Figures 14-4 and 14-5. Then the *segments* are formed by taking five macroblocks for each segment. The segments are assembled in a pseudo-random manner—within a segment the five macroblocks are taken from different positions in five different superblocks, no two in the same row or column of super blocks. At this stage, the segment consists of 30 blocks of 8x8 pixels, six blocks (four luminance blocks plus

Figure 14-2
Reduction to 4:1:1
sampling for 625/50
systems.

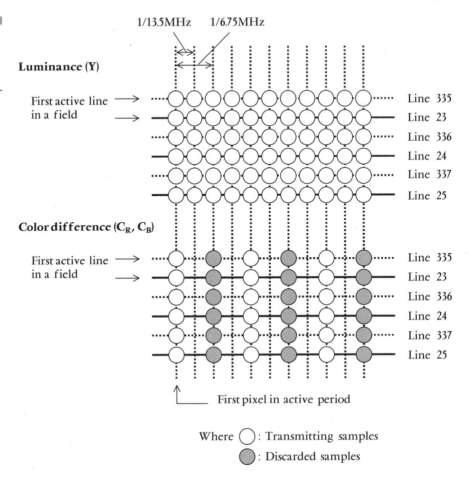

one of each color difference block) in each macroblock. Each pixel value is expressed as an 8-bit word, so the segment comprises 30×8×8 = 1920 bytes of data.

Each block is DCT transformed. DV provides for both frame and field DCT coding, but unlike MPEG-2, the decision is made for each block, and affects only that block. The two modes are known as *8-8-DCT*, used for blocks where there is little content variation between the two fields, and *2-4-8-DCT*, used where the content variation is significant (usually as a result of motion in that area of the image). In DV a 2-4-8-DCT is coded as two 8×4 blocks; the top block is the sum of adjacent rows of pixels, the bottom block the difference between adjacent rows. This is illustrated in Figure 14-6.

Figure 14-3
Arrangement of DCT
blocks for 4:1:1
encoding.

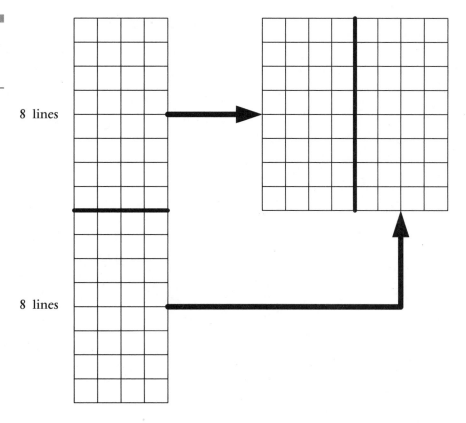

8 lines

8 lines

Luminance DCT block

Left   90 DCT blocks   Right

Top

Color difference DCT block   8x8 pixels

Left   22.5 DCT blocks   Right

Top

16x4 pixels

Figure 14-4

Superblocks and macroblocks in one television frame for 525/60 system with 4:1:1 compression.

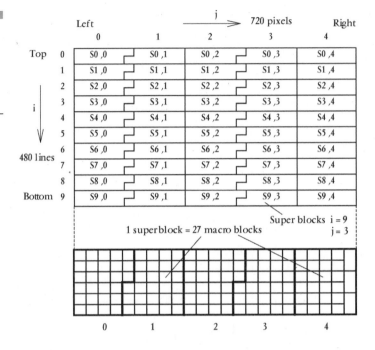

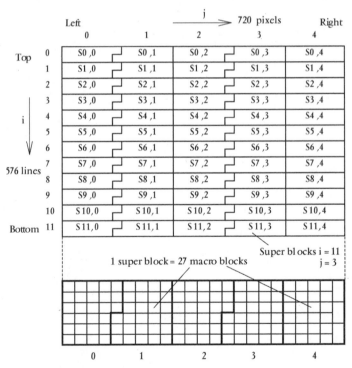

Figure 14-5

Superblocks and macroblocks in one television frame for 625/50 system with 4:1:1 compression.

Figure 14-6
The two DCT modes
of DV, and the
scanning order of the
resulting DCT
coefficients.

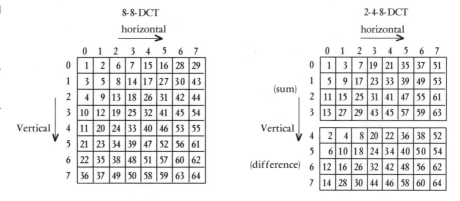

The DC coefficient is quantized to nine bits (-255 to $+255$), and initially the weighted AC coefficients are quantized to ten bits (-511 to $+511$). Each block is then allocated to one of four classes, labeled $0-3$, depending on the value of the largest AC coefficient in the block. Class 0 is used for the lowest energy blocks, class 3 for the highest. For class-3 blocks, the least significant bit of each AC coefficient value is discarded. Otherwise, the most significant digit of every AC coefficient is discarded. (Any AC coefficient greater than 255 immediately qualifies the block for class 3, so that for any other class the MSB of every coefficient must be zero.) At this stage both DC and AC coefficients are 9 bits.

Each 9-bit AC value is then divided by a quantization step, determined by the class number and an area number, dependent on the position of the coefficient in the transformed block. The area numbers are illustrated in Figure 14-7, and the derivation of the quantization step is shown in Figure 14-8.

Figure 14-7
Area numbers.

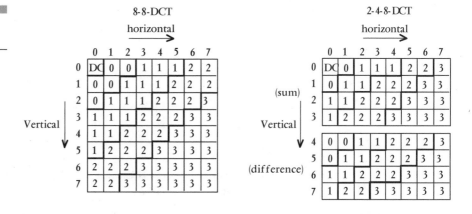

Figure 14-8
Derivation of
quantization steps.

	Class number				Area number			
	0	1	2	3	0	1	2	3
	15				1	1	1	1
	14				1	1	1	1
	13				1	1	1	1
	12	15			1	1	1	1
	11	14			1	1	1	1
	10	13		15	1	1	1	1
	9	12	15	14	1	1	1	1
	8	11	14	13	1	1	1	2
	7	10	13	12	1	1	2	2
Quantization	6	9	12	11	1	1	2	2
number	5	8	11	10	1	2	2	4
(QNO)	4	7	10	9	1	2	2	4
	3	6	9	8	2	2	4	4
	2	5	8	7	2	2	4	4
	1	4	7	6	2	4	4	8
	0	3	6	5	2	4	4	8
		2	5	4	4	4	8	8
		1	4	3	4	4	8	8
		0	3	2	4	8	8	16
			2	1	4	8	8	16
			1	0	8	8	16	16
			0		8	8	16	16

This table needs a little explanation. The *quantization number* is the compression control, and varies between 0 and 15. Zero represents the highest compression (coarser quantization of the coefficients, so fewer bits in the encoded output) and 15 the lowest compression. The system seeks the highest quantization number that can be used, consistent with the final output of the compression system being not more than 385 bytes for the complete segment. Each block in the segment has already been assigned to one of the four classes. Now the system can test each quantization level. Each AC coefficient in each block is divided by the appropriate quantization step. For example, for quantization number 7, in a class-2 block, coefficients in area 0 would be divided by 2, coefficients in areas 1 and 2 would be divided by 4, and those in area 3 by 8. For the same quantization number, coefficients in a class-1 block would be divided by 1, 2, 2, and 4 respectively.

It may appear that the table entries for class 3 are "out of place." This is because the AC coefficients of class-3 blocks have already been divided by 2 as a result of discarding the least significant bit.

The quantized coefficients are read out in the scanning order shown in Figure 14-6. The combination of run-length zeros and the following coefficient value is then variable-length encoded in a manner similar to JPEG and MPEG, but with code tables optimized for

DV. As in JPEG and MPEG a 4-bit *end-of-block* (EOB) code is inserted when there are no more non-zero coefficients. Note that the variable-length—encoding step must be inside the loop that measures the number of bytes required for each quantization number. In other words, for each quantization number, the coefficients must be quantized, scanned, and variable-length—encoded before the length of the encoded segment is known.

When the correct quantization number has been found (the highest number that results in not more than 385 bytes for the segment), the resulting data is arranged as shown in Figure 14-9, which shows the data for one macroblock. After a 4-byte header that identifies the macroblock, 77 bytes are provided for macroblock data ($5 \times 77 = 385$). The first byte carries 4 bits for the quantization number (0 to 15) and four bits to report error status and error concealment action in a videotape machine. Then comes the first luminance block DC coefficient (9 bits) plus one bit to indicate field or frame DCT, and two bits to indicate the class of the block (0 to 3). Then there is a space of $12\frac{1}{2}$ bytes for AC coefficients. The other five blocks of the macroblock follow in similar manner, except that only $8\frac{1}{2}$ bytes are provided for the AC coefficients of the color-difference blocks.

Figure 14-9
Arrangement of a compressed macroblock with 4:1:1 compression.

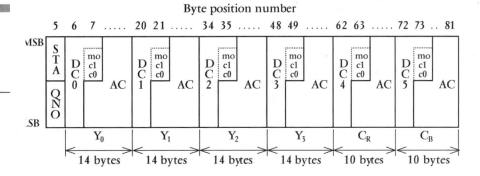

NOTES
STA: Error status
QNO: Quantization number
DC: DC component
AC: AC component
EOB: End of block (0110)
mo: DCT mode
c0, c1:   Class number

Of course, not all blocks will conveniently generate exactly $10\frac{1}{2}$ or $8\frac{1}{2}$ bytes of variable-length—coded AC coefficients. In some blocks the

EOB will occur before the end of the allocated space; for others there will be too much data for the space provided. DV employs a complex three-pass algorithm to take maximum advantage of the space available. The first pass places the data as described; the second and third passes attempt to place all the remaining data in the "empty" spaces after EOB in the blocks with little data.

So, for 4:1:1 compression of 525/60 there are 1,350 macroblocks per frame, or 270 segments per frame; approximately 30 frames/second. Each segment has a 4-byte header plus 385 bytes of data, so that the data rate is:

$$270 \times 389 \times 30 \times 8 = 25,207,200 \text{ bits/second}$$

For 625/50 systems with 576 active lines there are, 1,620 macroblocks or 324 segments per frame, and 25 frames per second, so the data rate is:

$$324 \times 389 \times 25 \times 8 = 25,207,200 \text{ bits/second}$$

Note that the 4:1:1 input data (at 8-bit precision) represents just under 125 Mbits/s, so that this mode of DV provides about 5:1 compression.

50 Mbits/s Compression

This section is much shorter because 50 Mbits/s compression uses all the techniques of the 25 Mbits/s compression just described. In fact, in practical implementations, the process uses two standard DV chips instead of one! 50 Mbits/s DV compression uses 4:2:2 video without discarding any pixel values and generates two streams of 25 Mbits/s data that may be multiplexed into one 50 Mbits/s stream. The differences lie in how the video data is distributed to the two channels and how the data is arranged. The 4:2:2 video data at 8-bit precision represents about 166 Mbits/s, so this mode of DV provides about 3.3:1 compression and is virtually transparent.

The macroblock for 4:2:2 DV compression has only four DCT blocks, two luminance and one each color difference, as shown in Figure 14-10. Again five macroblocks are grouped into one segment, and half the segments are sent to each channel.

With only four blocks per macroblock (and a compression factor of 3.3:1), quantization is less aggressive, and more space is needed for AC coefficients. This results in a revised layout for the compressed macroblock, as shown in Figure 14-11. As with the 25 Mbits/s system, a

Figure 14-10
Macroblock and DCT
blocks for 4:2:2
compression.

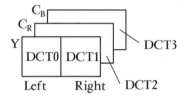

NOTES - DCTl: DCT block order
l = 0, 1, 2, 3

three-pass algorithm distributes excess coefficient data into otherwise
unused spaces in the data structure.

Figure 14-11
Arrangement of a
compressed
macroblock with
4:2:2 compression.

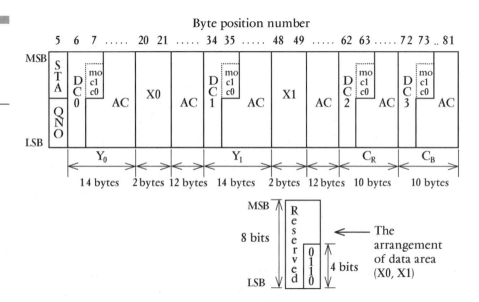

There are more differences in the detailed syntax of the bitstream,
but that is the essence of 4:2:2 50 Mbits/s compression in DV. All the
other processes described under the previous section apply.

Wavelets

Introduction

Wavelet technology offers the most viable alternative to DCT for image compression. The techniques were investigated by workers in many different fields, and existed under various names until the connection was spotted by a mathematician in the mid 1980s. Although the favorite child of theorists for many years, until recently practical image compression systems have failed to meet expectations.

The theory of wavelets is quite complex, but the principles involved are not. Like Fourier transforms and DCT, wavelets transform information into a representation using frequency-dependent coefficients, but wavelets differ in that useful positional information is retained. Before we look at wavelets, it will be useful to review some of the qualitative aspects of Fourier transforms.

More about Fourier Transforms

Let's look at the issue of frequency and location information for a moment because it's really important. If we have a series of samples of a signal, say in the horizontal direction, we can look at any sample and know immediately the amplitude information for that particular location in the horizontal direction. What do we know about the horizontal spatial frequency? Nothing! We need the information from many samples to gain much information about the frequency content of the signal, and from all of the samples to get all of the information about frequency content.

Now let's perform a Fourier transform. This gives us a number of values representing the frequency components of the signal. We can get the information pertaining to any given frequency by examining the single coefficient that represents that frequency. However, we have no information about where any amplitude event (such as a step) may occur. To find this we need to assess the contributions of all the frequency components—in fact, perform a inverse Fourier transform.

In general terms, in the spatial domain we have excellent location information but little frequency information; in the frequency domain we have excellent frequency information but little location information. Put like this it sounds very obvious, but it is a fundamental issue that is important to understand. One of the great things about wavelets is that they can give both frequency and location information.

It will help in understanding the operation of wavelets to examine in a little more detail how the Fourier transform works. The general form of a Fourier series for a periodic function is:

$$A_0 + A_1\sin(2\pi. f + \varphi_1) + A_2\sin(2\pi.2\,f + \varphi_2)$$
$$+ A_3\sin(2\pi.3\,f + \varphi_3) + \cdots$$

where A_0 is the DC component, f is the frequency of periodicity; *2f, 3f,* etc., are the multiples of the first frequency, and φ_1, φ_2, etc. are the phase angles of each of the component frequencies. Remember, as discussed in Chapter 5, Fourier transforms require two coefficients for each frequency; instead of using phase angles directly, we could have used one sine and one cosine term for each frequency, each with an appropriate coefficient. Let's examine how we arrive at these coefficients. Let's build a simple periodic waveform by adding a few Fourier terms:

$$f(x) = 3\sin(2\pi.x) + 5\sin(2\pi.3x) + 4\sin(2\pi.5x)$$

The waveform is shown in Figure 15-1. In this case I have made all the phase angles zero, so we need only worry about one coefficient for each frequency. The DC component is also zero.

Figure 15-1
A periodic wafveform made from three sine components.

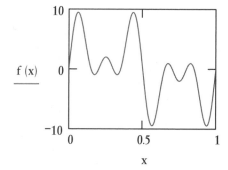

A Fourier coefficient may be calculated by multiplying the waveform by the corresponding Fourier component frequency and integrating the result over the period of the input waveform. (The DC coefficient is obtained by integrating the original waveform itself over the same period.) Normally we would have to use both sine and cosine components to get both coefficients, but in this case sine waves alone will be sufficient. So, if we multiply our sample waveform by the fundamental frequency (the frequency of periodicity), we have:

$$g(x) = 2.\sin(2\pi.x).[3\sin(2\pi.x) + 5\sin(2\pi.3x) + 4\sin(2\pi.5x)]$$

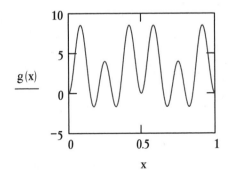

This product is shown in Figure 15-2. (There is probably a good mathematical reason for the multiplier "2" but I put it there to make the numbers come out right!) If we integrate this waveform over the full period:

$$\int_0^1 g(x)dx = 3$$

we have our first coefficient. Just to validate the simplification, if we change the sine to a cosine, the waveform product is shown in Figure 15-3, and the integral over the same period is zero. In fact, all cosine products will integrate to zero, so we can continue to look just at sine products.

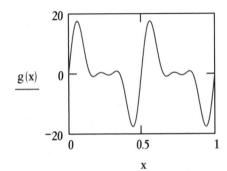

The behavior I am trying to demonstrate can be seen if we compare the results using the fourth and fifth harmonics of the fundamental. We expect to see a non-zero coefficient for the fifth harmonic, because that was part of our original waveform definition. There was no fourth harmonic in the original definition, so we expect to see a zero coefficient here. The result of the two multiplications is shown in Figures 15-4 and 15-5. In Figure 15-4 the original waveform is multiplied by the fourth harmonic. Because this frequency does not exist in the original waveform, the product is symmetrical and integrates to zero. In Figure 15-5 the fifth harmonic is used as a multiplier. Because this frequency was present in the original, we get a sine-squared component in the product. This component is always positive, so the product is asymmetrical and integrates to a nonzero result. In fact, the integral evaluates to 4, the coefficient value we would expect.

Figure 15-4
The fourth harmonic is not present in the waveform, so the product is symmetric, and we get a zero coefficient.

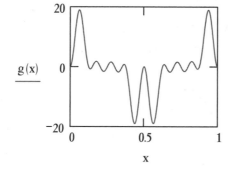

Figure 15-5
The fifth harmonic is present, so the sine-squared products are asymmetric; the integral and the resulting coefficient are non-zero.

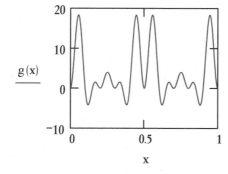

The behavior this demonstrates is that multiplying by a sine wave of given frequency "picks out" that frequency in the original waveform by creating a sine-squared product that does not integrate to zero. Any frequency not present in the original does not do this and (if it is a harmonic of the original's frequency of periodicity) creates a symmetrical product that integrates to zero.

Wavelets Concept

There are many ways of looking at wavelets, but one that fits with the concept we have just been discussing is to view wavelets as a burst of energy with a dominant frequency. A wavelet is shown in Figure 15-6. It is fairly obvious that if we were to multiply a signal by this waveform, the product would contain asymmetric information where the original signal had frequency content similar to that of the wavelet. Just as with the Fourier transform, integrating the product would give us a nonzero coefficient. Multiplication by the wavelet "picks out" detail from the signal.

Figure 15-6
A wavelet is a burst of energy.

That is obvious and interesting behavior, but two questions arise. Obviously, the result we get depends on the where on the signal we place the wavelet. How do we get information on the whole signal? Also, it would appear that a wavelet can pick out the detail in only one frequency range. We will address that issue a little later; meanwhile, let's look at the first question.

In reality, of course, we are not dealing with analog signals but with a sequence of digital samples. The wavelet does not really look like Figure 15-6, but more like Figure 15-7—a sequence of sample values, each of which will be used as a multiplier operating on a sample of the waveform. The particular wavelet shown is a sequence of 13 values, 6 either side of the origin. In applying a wavelet transform, this sequence is "walked across" the sample sequence (waveform) being transformed. This process is known as *convolution*, and is shown in Figure 15-8. For simplicity, this figure shows a short wavelet with just five

Figure 15-7
The wavelet of Figure 15-6 represented as a sequence of samples.

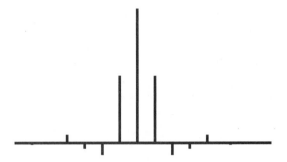

values, two either side of the origin. The wavelet is moved along, one sample at a time, and the convolution product evaluated for each position. For example, the wavelet is shown operating on sample S_7, and the output of the convolution is a new value S_7', where:

$$S_7' = W_{-2}.S_5 + W_{-1}.S_6 + W_0.S_7 + W_1.S_8 + W_2.S_9$$

Figure 15-8
Convolving a five-sample wavelet, W, with the samples of a signal, S.

				W_{-2}	W_{-1}	W_0	W_1	W_2						
S_1	S_2	S_3	S_4	S_5	S_6	S_7	S_8	S_9	S_{10}	S_{11}	S_{12}	S_{13}	S_{14}	S_{20} ...

One point that arises from this operation is that to convolve the wavelet over the whole sample set, such as an image, additional samples must be created at the edges. Typically an artificial extension of the image is created by a "reflecting" the edge samples far enough to accommodate the wavelet. In this example, only two additional samples are needed at each end of the waveform. At the left-hand side we create $S_{-1} = S_0$ and $S_{-2} = S_1$.

Wavelets as Filters

Some readers will recognize that the process just described is exactly the same as that shown in the Figure 15-9, which is a classical representation of digital filter. In this diagram, the samples of S are input sequentially to the left-hand side and pass through four delays, T, equal to the sample interval. The output, S_7', when sample S_7 is at the center of the filter, is the same as shown in the previous equation. The wavelet shape shown in Figure 15-6 is the impulse response of the filter shown in Figure 15-9.

Figure 15-9
The five-sample wavelet shown as a five-tap filter.

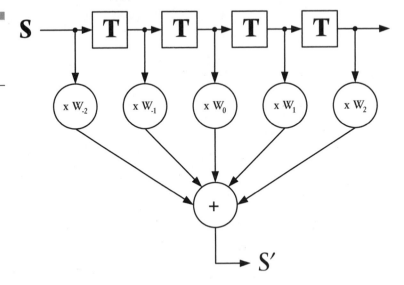

The idea of moving a wavelet over the image and picking out detail shows us how wavelets can give both frequency and location information. The concept of wavelets as filters gives us a better idea of how to use them for image compression. It also introduces another point—a wavelet is actually two complementary functions, or two complementary filters. One is known as the wavelet, the other as the *scaling function*. Although there are a great number of complexities in the design and implementation, the concept of a *fractal compressor* is quite simple, and in some ways quite familiar.

The image is filtered by convolving it with the wavelet, which behaves like a high-pass filter and extracts the high-frequency detail of the image. The image is also convolved with the complementary scaling function that removes the high frequencies. So now we have a set of wavelet coefficients representing the fine detail of the image, and an image from which the fine detail has been removed. It is possible to construct two-dimensional wavelets, but the transform is separable, so generally the two-dimensional image is handled by applying the wavelet and scaling functions in both the horizontal and vertical directions. The process is very similar to the *quadrature mirror filter*, described in more detail in Chapter 17. The image is split into two parts, high frequency and low frequency, operating, say, in the horizontal direction first. The two resulting subimages contain both high- and low-frequency vertical information. Each of the sub images is now convolved with the wavelet and the scaling function vertically, each producing two new separations. The process is shown in Figure 15-10.

Figure 15-10

Division of the image into four subimages.

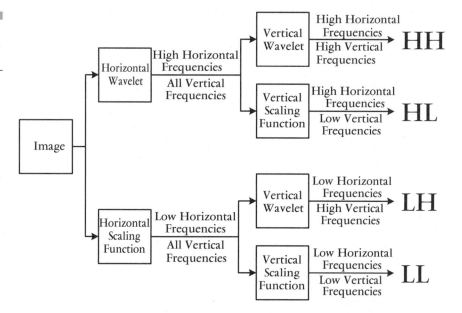

After this process, we have four subimages. One contains high horizontal and high vertical frequencies (HH), one has high horizontal and low vertical frequencies (HL); a third has low horizontal and high vertical frequencies (LH). Finally, the fourth has only low frequencies, horizontal and vertical (LL). Because of the bandwidth restrictions, the sampling density may be halved, horizontally and vertically, as explained in Chapter 17. Each subimage has, therefore, just one quarter of the samples or pixels of the original. This is shown in Figure 15-11(b).

It should be noted that the illustrations in Figure 15-11 exaggerate the amplitude of the wavelet coefficients. Each subimage is scaled to display the largest coefficient as white, to make visable as many coefficients as possible. This gives a better idea of the structure of the subimages, but a misleading impression of the energy of the high-frequency coefficients.

(a) The original image, as used elsewhere in this book.

This image is 256 x 256 pixels.

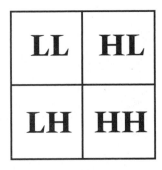

LL	HL
LH	HH

(b) A single pass of the transform, showing the three sets of coefficients, and the residual filtered image.

The amplitude of the coefficients is exaggerated by rescaling to make them visible.

Figure 15-11 Wavelet transforms of "Boats." (a) The original image, as used elsewhere in this book, at 256 × 256 pixels. (b) A single pass of the transform, showing the three sets of coefficients, and the residual filtered image. The amplitude of the coefficients is exaggerated by rescaling to make them visible.

(c) This example shows three iterations and the resulting ten subbands.

(d) After five iterations the residual image is reduced to just 64 pixels, shown above.

Figure 15-11   (c) This example shows three iterations and the resulting ten subbands. (d) After five iterations the residual image is reduced to just 64 pixels, shown above.

Wavelet Compression

Now we can start to put a system together. We left a question unanswered earlier—how do we deal with the fact that a wavelet appears to extract only one frequency range? In the previous exercise, we had the three subimages that contained high frequencies (HH, LH, and HL), plus the low-frequency subimage, LL. All of these subimages were reduced to half-sampling density. Now, suppose we convolve the original wavelet with the subimage LL. The wavelet is unchanged, but the image has been subsampled. A frequency that had, say, 8 samples per period in the original image now has 4 samples per period. Thus, the same wavelet captures frequencies at half of those captured in the first pass. Similarly, the unchanged scaling function now eliminates high frequencies down to half the cut-off frequency of the first pass.

Now we can see how the wavelet elegantly spans the void between spatial and frequency representations. Spatial representation gives excellent location information and no frequency information. Frequency representation gives excellent frequency information and no location information. On the first pass a wavelet gives coarse frequency information (the top half of the bandwidth) and fine spatial resolution (half the original sampling density). Each successive pass gives information about a smaller band of frequencies (finer frequency resolution), but with coarser location information, because of the subsampling that occurs on each pass. Eventually we would have good information about the lowest frequencies in the image, but with no positional information because we would have only one sample left!

Clearly we have an iterative process here. On each pass the image is split into four subimages, and on the subsequent pass the same process is applied to the quarter-size LL subimage, generating four sub-subimages! This process is easy to see in Figure 15-11(c), which shows three iterations. Figure 15-11(d) shows five iterations, and LL_5 is now reduced to eight pixels in each direction (from the original 256). We could, in theory, continue until LL had only one pixel (eight iterations with this image), but there comes a time when the overhead of identifying the different subimages exceeds the benefit of additional passes. A typical software compression application would use six passes on this image, leaving just 16 values from LL_6 to be transmitted.

This may seem like an enormous computational task. Certainly the convolution process is quite expensive, but we do not get the full impact on every pass because the image is smaller each time. If the first pass uses N processor cycles, the next uses $N/4$ cycles, the next $N/16$, and so on. For the six iterations discussed above, only about $4N/3$ cycles are used in total.

Wavelet-based compression is similar to DCT-based compression in yet another way. Nothing we have done yet as resulted in compression! However, the wavelet coefficients of an image have similar characteristics to the DCT coefficients. Many are zero or close to zero, and the remainder may be coarsely quantized with relatively little perceptual impact. We saw that the human psychovisual system permitted coarse quantization particularly of the high-frequency coefficients. The first pass of wavelet compression yields the largest number of coefficients (because the image is at its largest), and these are all high-frequency coefficients.

As with DCT-based compression, the next step is to organize the remaining coefficients to maximize the benefit of lossless compression techniques. In the discussion of JPEG we saw that the zigzag scanning "collected" the most significant coefficients early in the scan, and maximized the runs of zeros. These subimages resulting from the wavelet transform require more complex operations. One such technique is called *zerotree coding* and uses *significance maps*. This algorithm assumes that if a coefficient is insignificant in a low-resolution subimage, then the corresponding coefficients in the higher resolution subimages will also be insignificant. Further details are beyond the scope of this book, but the bibliography contains a wealth of references for those who wish to delve deeper.

So, wavelet-based compression and DCT-based compression are similar in many ways. Figure 15-12 shows a DCT-based compression system; Figure 15-13 shows a wavelet-based system. The only fundamental difference is the transform and, as we have seen, both DCT and wavelets are closely linked to Fourier transforms.

Wavelet compression does require more computational power than DCT-based compression. But, according to its proponents, it offers a number of substantial advantages.

■ Wavelet coding is inherently scalable. The transform process, shown in Figure 15-10, is repeated for as many iterations as required. The decoder performs the inverse process, *but may stop at anytime* if the full resolution of the original is not required. A bitstream may carry the full resolution, but a decoder may use a subset of the bitstream, depending on its capabilities and the resolution of the display being used.

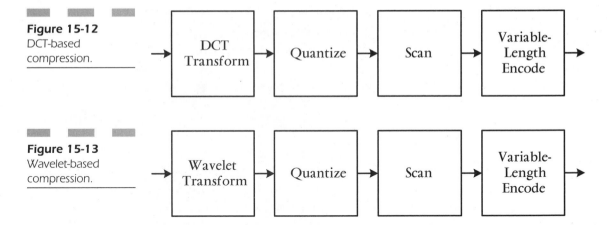

Figure 15-12
DCT-based compression.

Figure 15-13
Wavelet-based compression.

■ Wavelet compression is claimed to be more efficient at low bit rates. Probably the most significant element here is the fact that wavelet-generated artifacts are generally less objectionable than DCT-generated artifacts. When DCT is used at low bit rates, the excessive quantization tends to result in level differences between adjacent DCT blocks. These straight-line correlated errors are precisely the artifacts to which the human psychovisual system is most sensitive. Excessive quantization of wavelet coefficients leads to "smearing" of detail, but there is nothing in the system that will generate straight lines.

It is difficult to show a useful comparison between the two systems on the printed page, but Figure 15-14 serves to illustrate the qualitative differences. It shows the image "Boats" (originally 256 x 256 pixels, 64 kBytes) compressed by JPEG and by a wavelet-based compression system. In each case the compression ratio is about 17: 1, for a file size of just under 4 kBytes. The CD-ROM include some tools that allow experimenters to make their own comparisons.

At the time of writing, wavelet compression has made very little impact compared to DCT-based compression. There are number of reasons for this. Wavelets have been less successful than DCT-based systems in achieving good efficiency at the near-transparent compression ratios. Also, once DCT was adopted by MPEG, most development effort went into producing integrated circuits for MPEG— that is, DCT. Until recently, little specialist silicon was available for wavelet compression. However, this is changing, and now that wavelet compression has been adopted in MPEG-4 (for static textures) and in JPEG2000, wavelet implementations are likely to become much more common.

Figure 15-14
Compression of "Boats" at 17:1 by JPEG (above) and wavelets (below).

JPEG2000

Introduction

We discussed JPEG back in Chapter 7, and indeed both historically and as a source of the technology used in other systems, that is its logical place. The various iterations of MPEG, the development of DV, and the evolution of wavelet compression all followed JPEG. But the JPEG committee has not been idle. The JPEG2000 standard is now at the stage of Final Committee Draft and is expected to become an International Standard during 2001. JPEG2000, like MPEG, is a very rich standard, and this short chapter can offer only a brief summary and references to additional information.

Limitations of the Original JPEG System

The original JPEG was and is a powerful system. It provided the groundwork for the dominant compression systems in use today. When used in accordance with its design criteria it provides excellent results. Nevertheless, as we have seen in the course of this book, it does have its limitations. JPEG2000 set out to address all weaknesses of JPEG, so a good starting point is to summarize these weaknesses.

- JPEG is designed for continuous-tone images and performs badly on imagery with different characteristics.

- Specifically, JPEG is unsuitable for much computer-generated imagery, binary (black/white) images such as text, and compound documents with different types of content.

- JPEG has many modes, of which a large proportion are application specific. There is no universal decoder architecture.

- At low bit rates, JPEG artifacts, such as blocking, are very visible and restrict the usefulness of the standard.

- JPEG is very susceptible to transmission errors; typically even a small error rate results in substantial image degradation.

- JPEG has no convenient mechanism for handling very large images.

- JPEG is inflexible by today's standards. It cannot support lossy and lossless compression in the same bitstream, and there is no ability to encode the important parts of an image to a higher quality than the rest.

Goals of JPEG2000

All these factors were taken into account in specifying the requirements for the new JPEG2000 standard. These include:

- Backward compatibility with JPEG; a JPEG2000 decoder should be capable of decoding a JPEG bitstream.
- Superior low bit-rate performance; initial tests show 20 to 50 percent reduction in bit rate for comparable quality, depending on image content.
- Ability to process large and/or high-precision images. JPEG2000 permits image height and width up to about 4 billion pixels. There can be up to 255 components in each image (e.g., Y, U, V, alpha, etc.) and each component may have any depth from 1 to 32 bits.
- Continuous-tone and bilevel compression.
- Lossless and lossy compression. JPEG2000 provides wavelets like those discussed in the previous chapter, plus wavelets with integer coefficients to permit lossless transforms. It also provides a special form of color difference coding using integer coefficients to move between *RGB* and *YUV*, permitting perfect reversibility.
- Progressive transmission, providing the gradual buildup of an image by resolution (spatial scalability) and by pixel accuracy (signal-to-noise scalability). Also, the ability to extract a lower-resolution image from the bitstream by a less-powerful decoder.
- Random access to the bitstream (start decoding at any point).
- Robustness in the presence of bit errors.
- Region of interest (ROI) coding and decoding, permitting higher quality for the most important parts of the image. Regions can be defined in a manner similar to shape coding in MPEG-4.
- Content-based description and image security, including encryption capability, indelible copyright information, and the like.
- Provision of an *alpha* or transparency channel.

JPEG 2000 is based on wavelet compression and offers a wide range of lossy and lossless filter combinations. At the time of writing most available software is experimental, but some excellent results may be demonstrated. Figure 16-1 shows a comparison between the conventional JPEG implementation of a popular graphics program, and an experimental JPEG2000 package. The compression is about 60:1 in each case (the original file is 64kB, the JPEG files is 1,361 bytes, and the JPEG2000 file is 1,305 bytes.

Figure 16-1
JPEG (above) and
JPEG2000 (below)
compression of
"Boats" by about
60:1.

Audio
Compression

Introduction

This will be a short chapter, partly because of limited knowledge on my part, but also because audio compression is built on many of the same principles as video compression. Only in one sense is audio compression radically different from video compression—it is much older and has been commercially exploited in the analog domain. It was not called *compression*; it was known as *noise reduction*, but essentially these are the same. A noise reducer modifies the signal such that when reverse processing is applied after transmission or recording, the noise of the transmission or recording channel is attenuated. The well known Dolby A, B, and C systems are actually simple, analog, subband companding systems—terms we encounter below.

Nevertheless, our interest is on compressing digital audio. Depending on the quality requirements, audio may be sampled at a variety of frequencies, usually in the range 8 to 48 kHz, and with sample resolutions from 8 to 24 bits. The lower limits provide telephone-grade audio; the higher can match or exceed human hearing capabilities. Of most interest in the professional world are two standards at the high end. Compact disc audio is sampled at 44.1 kHz, with a word length of 16 bits for each of two channels, a total of just over 1.4 Mb/s. Professional studio audio is usually sampled at 48 kHz with a word length of 20 bits, almost 1 Mb/s per channel. This rate can accommodate audio frequencies to over 20 kHz with a dynamic range of 120 dB.

As with any form of compression, the objective is to reduce the data by eliminating redundancy. There is "absolute" redundancy—data can be removed by an encoder then unambiguously recreated by a decoder. Removal of this redundancy is lossless compression. There is also perceptual redundancy—data can be removed without significant change to the experience of a human observer.

Perceptual redundancy itself falls into two categories. There are phenomena to which the observer is intrinsically insensitive. In video we saw that the higher-frequency transform coefficients could be coarsely quantized without significant impact on the viewer. This is true for all conditions. Another effect, used in bit allocation in video systems, is masking. An artifact that might be visible and objectionable in an image area with little activity (empty sky, for example), becomes invisible or insignificant in a busy area of an image.

All these effects exist in audio, and all are exploited by compression systems, but masking is by far the most significant contributor to data savings in a single-channel audio system. (Multiple channel systems also benefit from correlation between the various channels.) To understand the mechanisms of masking in the human hearing process we must examine the workings of the inner ear.

Masking in Human Hearing

The human hearing system has remarkable performance. Its useful range covers about 10 octaves and, in the more sensitive regions, it provides a dynamic range of more than 100 dB. The discriminating power of the ear—brain combination is incredible; we can resolve conversation and other intelligence from noise levels and interfering signals to a degree that seems to defy the laws of physics. This performance requires some very sophisticated processing.

One of the key elements of the ear is the basilar membrane of the inner ear. It separates two of the fluid-filled chambers of the cochlea and is in contact with the hairs of the organ of Corti. These hairs drive the actual sensors that send nerve messages to the brain. The basilar membrane varies in width, thickness, and rigidity along its length. This makes it frequency sensitive; different areas vibrate at different frequencies. The basilar is not just a passive membrane; it has an active mechanism that can provide positive feedback to low-amplitude vibrations. The combination of these characteristics leads to interesting behavior. (This explanation is somewhat simplistic. Various authorities have suggested different mechanisms for the behavior of the inner ear, particularly in respect to the active attributes. Some suggest nonlinearity in the cochlea; others attribute the active element to the organ of Corti. Suffice it to say that the behavior is complex, certainly not linear, and that there are active elements within the inner ear. We concentrate on the effects of this complexity.)

Some authorities estimate that there are 24 regions in the basilar membrane; others suggest a higher number, but it is certainly finite. Each region can vibrate over a small range of frequencies, but at only one frequency at a time within that range. When positive feedback is applied, the effect is that of a very high-Q tuned circuit. Each region vibrates at a frequency determined by the strongest stimulus within its range, *and is unaffected by any smaller stimuli*. This means that within each frequency band, only the loudest frequency contributes to the signal received by the brain, and so to what we hear.

This effect is the single most significant contributor to our ability to compress audio.

The effect is called *frequency masking*, and it contributes to our ability to compress in two distinct ways. As mentioned above, only the strongest stimulus within a region is significant. Any frequency in the signal in the same band, but lower in amplitude, need not be coded. Also, noise within the band is irrelevant, provided it is sufficiently below the prime stimulus. This allows encoding of the prime stimulus with a relatively small number of bits; the resulting quantization noise is masked just like any other noise. This effect is quite remarkable and, thanks to Dolby Laboratories, there is a demonstration on the CD-ROM that is well worth trying.

The positive feedback of the basilar contributes to another effect, again similar to that of a high-Q tuned circuit. The vibration responds slowly to changes in the amplitude of the stimulus, and this results in *temporal masking*—we fail to hear sounds a short time before, and a longer time after, a strong stimulus.

Simple Audio Compression Schemes

Many of today's consumer devices use some form of audio compression. Sometimes this is used to increase storage capacity. One example is found in 8-mm video recorders; compression is employed to reduce the tape area needed for audio. Audio workstations become much more cost effective if the capacity of the hard drives can be increased by using compression, provided the necessary high quality can be maintained.

In some applications, the compression is used for more than one purpose. The mini compact disc achieves its recording density by means of compression, but data reduction is also used to improve replay under adverse conditions. Data is read from the mini CD at the same speed as from a regular CD—about twice as fast as necessary for normal replay. The data enters a buffer, and after the buffer delay (typically a few seconds) it is decompressed and presented to the listener. If the listener is an energetic jogger and jars the player enough to move the read head, the player can continue to play from the buffer while the head is repositioned, and the additional data speed allows the buffer to be filled again, ready for the next "jog."

Simple schemes such as these generally use some form of *companding*. The number of bits required by each word is reduced by dynamically altering the level or gain according to the instantaneous amplitude, or the amplitude range of a group of samples.

Video 8, for example, uses a simple four-step gain control based on instantaneous signal amplitude. Low amplitudes are passed at unity gain; higher amplitudes are gain-reduced by factors of 2, 4, or 8. The result looks to a video engineer like a gamma curve (see Figure 17-1). The higher levels are coarsely quantized, but the increased quantization noise is masked by the high signal level. (In the case of Video 8, this mild reduction in data rate is supplemented by an analog compressor.)

Figure 17-1

Transfer function for Video 8 digital compression.

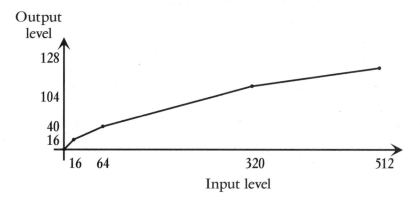

More sophisticated companding systems usually divide the input data stream into frames, typically between 1 and 32 ms long. In a 16-bit system the values for each sample can range from −32,768 to +32,767. However, it is rare for the sound in one frame to occupy the full range of sample values. If it does not, each sample can be expressed by a shorter word, combined with a gain or scaling factor applied to the complete frame. In other words, the levels are expressed in floating-point form, but with the restriction that the mantissa must have the same value throughout a frame. The NICAM systems, as used in stereo television, use this technique to reduce 14-bit words to 10-bit words, plus a 3-bit mantissa for each frame of 32 words.

There are two real problems with such systems. A frame with a single high sample value can cause low-level audio to be coarsely quantized. This is not usually audible if the frame length is short, typically not more than 1 millisecond. The greater problem is that they do not provide the large compression ratios we would like to see for many applications.

Quadrature Mirror Filters

To make efficient use of the masking effects of the human hearing system, it is necessary to split the audio band into regions as small as or smaller than the regions of the basilar membrane's response. Most modern audio compression schemes are based upon some system that divides the audio band in this way. We will examine subband systems, where the division is performed with filters, and transform systems, which use techniques more like those we have seen in video systems. Before we explore subband systems, we must learn about an important trick, the *quadrature mirror filter* (QMF).

We have decided that we need to split the audio spectrum into bands to benefit from the behavior of the inner ear. However, by conventional wisdom, band splitting means a large increase in bit rate if every band requires a sampling frequency of twice the highest frequency within the band. Common sense tells us there should be a better answer than this—we are not increasing the amount of information being sampled, so there should be a way of sampling that does not increase the total bit rate.

The trouble with low sample rates, as we saw in Chapter 2, is aliasing. Figure 17-2 reminds us how this occurs. However, if we are splitting a frequency band, we do not need the full spectrum shown in this drawing. Let's consider a simple example where we want to split the band into two. For ease of nomenclature we'll call the full spectrum 0 to $2f_a$, which we want to split into two bands, 0 to f_a and f_a to $2f_a$. To sample the whole spectrum, we need a sampling frequency of twice $2f_a$, or $4f_a$ (Figure 17-3). Now let's split the band as shown in Figure 17-4. We now have two half-band signals, each sampled at the $4f_a$ rate, so that we have twice as much sampled data as we started with.

Figure 17-2
Frequencies above half the sampling frequency cause baseband and sideband overlap, resulting in aliasing.

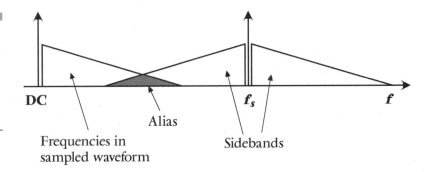

DC f_s f

Alias

Frequencies in sampled waveform

Sidebands

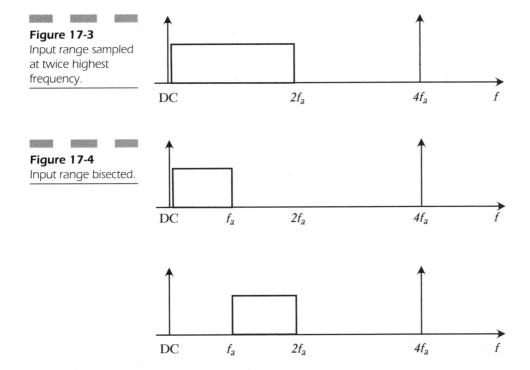

Figure 17-3
Input range sampled at twice highest frequency.

Figure 17-4
Input range bisected.

The lower part of the band is easy to deal with; we can discard every other sample, reducing the sampling frequency to $2f_a$ without risk of aliasing. What can we do about the other half-band? Thinking back to analog techniques, we could heterodyne the upper half-band down to the same frequency range as the lower band by mixing with a suitable frequency and filtering the result. We could then sample the resultant at $2f_a$, and after reconstruction heterodyne back to the original frequency range (Figure 17-5).

In fact there is a much simpler mechanism. Again, we'll repeat a drawing from Chapter 2, showing how the signal and its alias are indistinguishable because they have identical samples (Figure 17-6). But if we eliminate the lower half of the band, *only* the "alias" gets sampled, so there is nothing to confuse with it! All we need to do is use a bandpass reconstruction filter *that will generate only frequencies in the upper half-band,* and we have an unambiguous situation again. So, in fact, we can again discard alternate samples and the upper half-band also has an effective sampling frequency of $2f_a$ (Figure 17-7). We have succeeded in splitting the band while maintaining the same total number of samples.

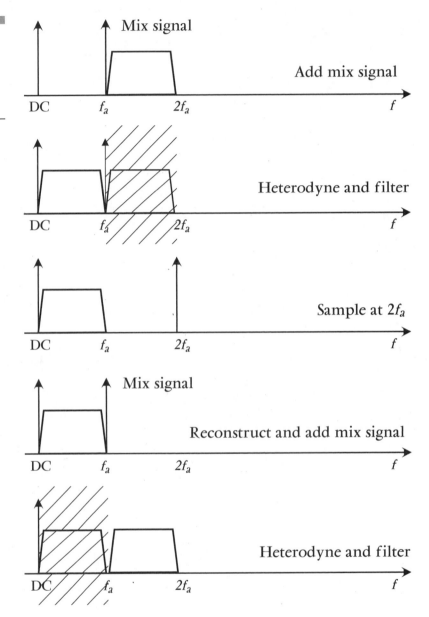

Mix signal

Add mix signal

Heterodyne and filter

Sample at $2f_a$

Mix signal

Reconstruct and add mix signal

Heterodyne and filter

▬▬ ▬▬ ▬▬

Figure 17-6
The same set of samples can represent either signal. Normally the lower frequency is in the valid (Nyquist sampled) band, but if this band is not present the samples unambiguously represent the higher frequency.

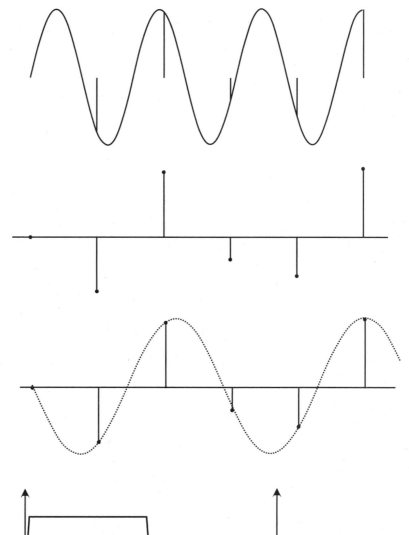

▬▬ ▬▬ ▬▬

Figure 17-7
The band is split without increasing the total data.

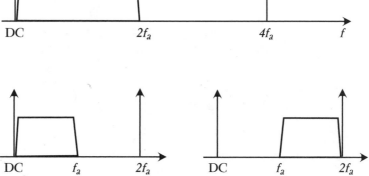

Having done this once, we can do it again. In fact, we can split the frequency band into any number of binary-related subbands without increasing the total amount of data. Figure 17-8 shows a nested arrangement of QMFs that could be used for this purpose. This arrangement splits a band 0 to f, sampled at $2f$, into eight subbands, each sampled at $f/4$.

Figure 17-8
Using a tree of quadrature mirror filters (QMF) to split a frequency band into eight equal subbands.

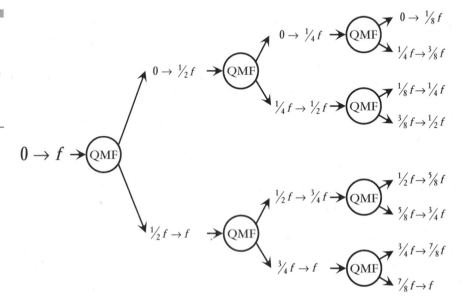

As John Watkinson (1995) points out, this arrangement calls for the computation of a large number of samples that are subsequently discarded. It is possible to build a filter that produces, say, 32 subbands and uses multiple coefficient tables that calculate only the needed samples. However, this process is mathematically and electrically so close to the computation of a transform that it is arguable what it should be called.

Now that we have demonstrated the ability to split the audio spectrum into bands without increasing the amount of data, we can now look at how to reduce the data in each band.

Subband Coding

Like many of the techniques we have encountered, subband coding is a means of arranging the signal data in such a manner that the tools we have available work most efficiently.

The principal tool for audio compression is companding. In simple companding systems we rely on the total signal lying within a limited range. If we split the audio into subbands it is much more likely that, within each band, we will encounter a limited range—in fact some bands may have no signal at all. Furthermore, we have seen that if the band matches, or is smaller than, a critical band of the human hearing system, masking may mean that there is very little that we need to code. Many subband coders use 32 subbands, each approximately one-third of an octave wide.

Before going further in this direction, we need to look at masking a little more closely. Like most functions associated with the operation of the human senses, masking is nonlinear and asymmetrical. In particular, a stimulus will have a substantial masking effect at higher frequencies, and at high stimulus amplitudes this effect can extend for several octaves. At lower stimulus levels the effect is smaller and extends for less than an octave. At all stimulus levels, there is very little masking of lower frequencies. When coding within a band, therefore, we must consider the worst case where the largest stimulus within that band is at the top edge of the band. Within each band, a threshold can be determined from the level of the largest stimulus. For reasonably high stimulus levels, and one-third octave bands, the threshold might be 20 to 30 dB below the stimulus. Nothing, including noise, below this threshold will be perceived, so that we can allow the quantization noise to approach this value; this means we can code the samples in that frame in that subband with 4 or 5 bits per sample! This is enough to convey the essential information within that narrow band. The presence of that information ensures that we will not hear the quantization noise, nor will we miss any smaller signals that might have been present in the original.

This is really a two-step process. On the basis of the highest level in the frame, a scale factor that moves this level to near the top of the coding range is applied to all samples. On the basis of the actual level, the masking threshold is determined, and the modified samples are truncated to this number of bits. The scale factor is sent with the frame to correct the gain in the decoder.

If there is a large signal in one band, the masking effect of this may mean that several higher bands may be encoded with even fewer bits, or not encoded at all.

Bit Allocation

All the above assumes we will always operate on the edge of audible errors. Rapid change of signal amplitude within a frame, not to mention the variability of human ears, means that a practical system should strive for some headroom above the theoretical threshold in each subband. Compromises may have to be made, because a practical system may be required to deliver a constant bit rate, irrespective of the input signal. The compression system needs a controller that allocates bits to the different subbands, based upon the needs of each band, the overall availability of bits, and its psychoacoustic model of human hearing. This controller may operate by analyzing the data in the subbands, or it may use a Fourier (or similar) transform of the input signal that can give a more detailed analysis of the input.

Whatever decisions are made in allocating bits must also be known at the decoder. There are three possible strategies.

In a *forward adaptive* system, the coder performs all calculations, and the allocation is encoded and sent to the decoder. This strategy has an advantage in that the encoder algorithms may be updated without affecting decoder operation, but some part of the bandwidth must be used to send the bit allocations to the decoder.

Backward adaptive systems perform the same calculations in the encoder and decoder. The decoder will always reach the same conclusions as the encoder, so that no allocation data needs to be sent. However, the decoder cost is significantly higher, and the encoder cannot be changed.

A compromise is found in *forward/backward adaptive* systems. Complex calculations are performed at the encoder, and a very small transmission bandwidth is used to send a few key parameters to the decoder, which then need only perform simple calculations. As in the backward adaptive approach, the encoder cannot be significantly changed, but some parameters can be changed to adjust behavior.

Transform Coding

In studying subband coding, we see that a number of factors point to the advantage of narrow subbands. (Remember, we have shown that we can split the input signal into any number of bands without increasing the data.)

- A typical audio spectrum is made up of many discrete frequencies. With wide subbands, most or all subbands will contain one or more components, and so have to be encoded. With narrower bands, more subbands will fall within gaps in the spectrum and contain no components. These bands do not have to be encoded at all.

- The number of bits needed to encode a subband depends on the degree of masking. We must always consider the worst case, where the stimulus is at the top edge of the band, because of the steep slope of the masking curve below the stimulus frequency. For this reason, narrow bands mask much higher noise levels and can be encoded with fewer bits.

Band-splitting filters are an efficient means of obtaining a reasonable number of subbands, such as the 32 used by many subband encoders. Beyond this, however, use of filters becomes very cumbersome, and a frequency transform is used to analyze the signal. Very large numbers of subbands may be obtained by this method. A 256-subband transform filter has approximately the same complexity as a conventional filter for 32 subbands.

There is a drawback to transform filters. Filters with high-frequency resolution have poor temporal resolution, and this affects the coding of frames that include transients. If the full coding gain is taken, the resulting quantization noise is present throughout the frame. A transient in the frame will probably mask this noise after the transient, but there is very little "negative" temporal masking; thus, the noise may be heard at the beginning of the frame.

To counter this effect, some systems use transient detection and switch to a lower number of bands for frames that include a transient. A correctly designed transform filter bank may be switched quite easily without any significant increase in complexity.

Example Compression Systems

Audio Compression in MPEG

MPEG has standardized on compression schemes based on the MUSICAM system developed by Philips. Three coding layers are defined for MPEG-1, but Layer II is by far the most used.

Layer II supports audio sampled at 32 kHz, 44.1 kHz, and 48 kHz. It uses 24-ms frames (with 48-kHz sampling) and codes 32 equal subbands. For each subband a 6-bit scale factor is used, giving a range of 120 dB. The scale factor can operate over the complete frame, but can be changed on 8-ms boundaries if required.

This system uses forward adaptive bit allocation and fractional bit quantization. It can code mono or stereo at bit rates of 32 to 384 kb/s. Layer II has been adopted for several direct broadcast satellite (DBS) systems and by the European Digital Video Broadcasting (DVB) system. It is also the standard for multimedia applications such as CD-ROMs.

MPEG-2 added some flexibility to MPEG-1 audio. Half-sample rates were added (16 kHz, 22.05 kHz, and 24 kHz) and provision was made for a multichannel extension. This approach provides backward compatibility with MPEG-1. If multichannel sound is required, a stereo mix must be derived and transmitted as Layer II audio. The additional information is sent in an enhancement stream. Multichannel decoders accept both streams, and remultiplex for the required number of channels. Simple decoders receive only the stereo bit stream and ignore the enhancement layer. Total bit rates of up to 1 Mbit/s are permitted for highest-quality multichannel.

The most well-known audio compression system is, of course, MP3—the designation for MPEG Audio Layer III. This uses more sophisticated algorithms to achieve greater compression efficiency, and is claimed to offer CD quality at about 64 kbits/s per channel. The availability of many thousands of MP3 compressed files on the Internet has done more to change the way we think about entertainment content than any other single development. It is far from certain how the copyright issues will be solved, and who will win the battle for control of digital entertainment distribution, but it is certain that the new techniques are here to stay.

Audio Compression for ATSC

During the process leading to the U.S. ATSC standard, it was decided to use MPEG-2 (MP@HL) for video coding. However, no agreement could be reached on the audio compression system to be used.

One proponent offered a system where six channels were individually compressed, but the main contenders were Philips with an enhanced MUSICAM system and Dolby with its AC-3 system. Both systems offered "5.1" channels. This means five full-bandwidth channels for front left, front center, front right, left surround, and right surround. An additional channel has only narrow bandwidth and is intended to drive subwoofers, chair-shakers, and other very low-frequency devices.. It is also known as the *low-frequency effects* (LFE) channel.

In a bitterly fought contest, expert listeners compared the two systems under carefully controlled studio conditions. AC-3 was selected, on the basis of its slightly superior performance on certain vibraphone passages. Philips contested the decision on the grounds that the MUSICAM system had not been performing correctly at the tests, but the decision remained. In fairness it should be said that when either system is working optimally, expert listeners cannot distinguish the results from the original recording, although Dolby claims this performance level at a lower bit rate. AC-3 supports bit rates from 32 to 640 kbits/s; the ATSC tests were performed at the intended transmission rate of 384 kbits/s (the maximum permissible rate has now been increased to 448 kbits/s, the same maximum as specified for DVDs). This represents a compression ratio of about 10:1, making the quality a remarkable achievement.

AC-3 is a transform system, switching between 256 subbands for normal operation and 128 subbands for transients. It supports the usual sampling rates of 32 kHz, 44.1 kHz, and 48 kHz, and uses 32-ms framing at 48 kHz. It uses a 4.5-bit scale factor over a range of 144 dB and forward/backward adaptive bit allocation.

Dolby incorporated a number of interesting features in AC-3, to be known as Dolby Digital in the commercial world. Although the system can encode up to 5.1 channels, the decoder can produce whatever downmix is appropriate for the receiver capabilities. Decoders can downmix to mono or to stereo or, of course, support full 5.1-channel output. In addition, each Dolby Digital (DD) decoder incorporates a Dolby Surround Pro Logic (DSPL) decoder to support the most common surround-sound installations. In fact, Dolby Digital includes a specific pass-through mode for DSPL encoded audio. The two DSPL signals are carried on the left and right front channels of the 5.1 system and passed directly to the DSPL decoder.

It is important to recognize the existence of this mode. Many videotapes, particularly film-tape transfers, have DSPL audio. DSPL is an analog compression scheme that carries center and surround information coded into the left and right channels; it is a four- to two- to four-channel system. It might be assumed that a DSPL signal could be decoded to four channels and fed to a DD system, perhaps feeding the surround signal to both left and right rear channels. This would work if there were five loudspeakers at the receiver, but there is a hidden trap.

DSPL is an analog system where the matrixing can never be perfect, so the system is designed with this in mind. Specifically, the system is designed to avoid sensitivity to phase variations in high frequencies in the surround channel. Also, DSPL is not a true surround-sound system in that it does not attempt to place sounds accurately in a 360° stage. It provides accurate location in front of the listener, plus a general (and very effective) ambiance from behind. As a part of this process the DSPL decoder low-pass—filters and delays the derived ambiance signal. Provided the four signals are then conveyed transparently to four loudspeakers, all will be well.

However, if the DSPL signal is decoded and fed to a DD system and decoded at a receiver with only (two-channel) stereo capability, the DD decoder downmixes for the stereo output. If the signals it mixes include the filtered and delayed surround from a DSPL decoder, the downmix will be quite wrong and objectionable.

To avoid this problem, DD provides the DSPL mode described above. The two DSPL-encoded channels are fed to the left and right front channels of the DD system, and the "message" that this is DSPL is passed with the bit stream to the decoder, which passes the two channels directly to the incorporated DSPL decoder. No downmix is required for stereo, and the two DSPL channels can be combined for mono.

Dolby Digital also addresses the perennial problem of loudness variation between programs and between channels. The system uses dialog level to normalize receiver volume; the bit stream carries an indication of the dialog level and the decoder gain is varied to keep this essentially constant.

Finally, DD allows for adjustable dynamic range at the decoder. Wide dynamic range is perfect for home theater environments, but quite unsuitable for a portable television in a noisy kitchen. The decoder can reduce dynamic range as determined by the receiver manufacturer—or according to the preferences of the listener if user controls are provided—with separate control of compression above and below the nominal dialog level.

Some issues surrounding the use of DD for digital television have not yet been resolved. One is the metadata (descriptive data) that should accompany the audio to the encoder. The encoder must know if its input is DSPL, so as to invoke the pass-through mode. It also needs to know the nominal dialog level and the dynamic range setting in order to convey these to the receiver. This presents a practical issue in a television plant where no obvious mechanism exists for routing, recording, and distributing the metadata with the audio.

Initially most potential users assumed that DD would be used for network distribution as well as final transmission to the home. However, fades, voiceovers, and the like. cannot be performed on the coded signal, and such operations in a local TV station require decoding and re-encoding. As with any compression system, this results in quality loss. It has been suggested that DD coded at a higher rate than that used for transmission could be used for distribution to provide the necessary quality headroom. However, it is possible that a different coding system, Dolby-E, may be preferable for distribution; this is discussed briefly in Chapter 19 under Mezzanine Compression Systems. As with many aspects of digital television, implementation is very slow, and new decisions are being made all the time.

Streaming Media

Introduction

Streaming is a technique used to deliver multimedia content, often including video and audio, usually compressed. It is used on networks, including private intranets and the Internet. Streaming brings together characteristics of two fundamentally different delivery systems.

In the world of digital studio installations, both video and audio, connections between pieces of equipment are usually made by dedicated circuits. These circuits carry the bitstream, usually uncompressed, in real time with no significant transit delay. They are deterministic and isochronous—bits go in to the circuit at a fixed rate and bits come out at the same fixed rate. For every bit that goes in, one comes out. The data is received by hardware elements designed specifically for the communication circuit being used, and every usable element, be it an audio sample or a video pixel, is processed as it arrives.

In this scenario there is really no concept of a beginning or end; the bitstream may last for a second, or for many years. The transmitting device sends the information it has; the receiving device accepts the information as it receives it. Even when there is no real information to send—if the video is black or the audio silent—as long as the equipment is powered, it will faithfully send the codes for black or silence, and they will be diligently received.

In the world of computers, the usual method of transferring information is a file transfer. One application creates a file and causes it to be stored on some medium. Another application may access the file from its original storage location, or some separate mechanism may cause the file to be transferred to a different storage location before the second application accesses it. The transfer mechanism may move a physical medium like a floppy disk ("sneakernet"), but today it is more likely that the file is transferred over some form of network. The file may be transferred quickly or slowly, perhaps in many parts, and the parts may not arrive in the same order as they were sent. It is the job of the transfer mechanism to ensure that when the job is finished, the received file is the same as the original.

Whatever the mechanism, the process is essentially serial. The file is created, then moved, then accessed. If there is no information to send, no file is created. No access to a file is possible until the transfer is complete.

This is the mechanism used, for example, when downloading an MP3 file from a music distributor. The file is transferred over the Internet from the server to the user's hard drive. Only when the transfer is complete can the file be accessed by an MP3 player.

The first system described above is really a very specialized form of streaming, but the term today is generally used to refer to transmission of content over a network. What makes streamed data special is that the content may be accessed and used without waiting for the completion of a file transfer. An example of streaming is the sample audio clips offered by on-line vendors of CDs. Many offer the ability to hear a few seconds of one or more tracks before making a purchase decision.

Streaming in this environment is not the same as the isochronous studio interface. We must assume a connection that supports a rate at least close to the data rate of the clip being played. If the clip is encoded at 384 kbits/s and we attempt to stream over a 20 kbits/s link, the best we could offer is one second of audio followed by a 20-second wait, then one more second of audio. In this case, waiting for a complete file transfer and playing the clip from the hard disk is the only practical solution.

But even if the average link speed is sufficient to support approximately real-time transfer, the transfer rate will rarely be constant. Delivery via the Internet generally involves many links and many routers. Transmission is usually "bursty" and packets may not be delivered in the order that they were transmitted. Clearly buffering is needed to permit an acceptable presentation. An intelligent player application will adjust the buffer time according to the characteristics of the incoming data; if the buffer empties (resulting in a break in presentation) the buffer size will be increased to make future interruptions less likely.

Applications for Streaming Media

The usefulness of streaming media is totally dependent on the availability of adequate bandwidth. The comments that follow are based on the environment as seen late in the year 2000. I make some pessimistic predictions below, and I hope I shall be proved wrong, as most commentators are! Time will tell.

In corporate intranets this may not be an issue: 100 Mbits/s Ethernet is starting to replace 10 Mbits/s, star network topology is rapidly replacing loops, and switches are replacing hubs. All of these factors help to increase the bandwidth available to a user and to reduce bottlenecks in the network. Streaming video over corporate networks is viable today, and the situation is continually improving.

Away from this environment the situation is much less encouraging. Audio streaming is useful on the Internet. The standard home connection is a 56 kbits/s modem, often achieving about 40 kbits/s. Modern techniques of audio compression can offer acceptable speech quality at 10 kbits/s or less and reasonable music quality from about 20 kbits/s. The standard connection, and the Internet itself, offer sufficient overhead for these rates to be supported quite consistently.

As mentioned above, e-commerce vendors can offer music samples that genuinely improve the shopping experience. Congressional hearings and shareholder meetings are now routinely "broadcast" over the Internet.

Video is a very different story and has, to date, achieved little more than novelty value over the Internet. High-quality, full-screen, full—refresh-rate video still requires megabits per second. Improvements in compression technology will help, but there is no way we will be able to deliver serious video in a few hundred kilobits per second. While we are confined to connections with even lower bandwidths, Internet streaming video will likely remain postage-stamp sized, and/or have very a low refresh rate.

There are useful applications for streaming video. As mentioned above, corporate intranets can provide a suitable infrastructure. Aside from this, video conferencing and distance learning are attractive applications that can deliver real benefits even today, but the value is greatest where the video is a supplement to other content (audio, faxed documents, prepublished hard copy material, electronic slide presentations, etc.). Generally video conferencing and distance learning use leased network connections or dial-up ISDN, offering at least 128 kbits/s. Even at this, quality is low, update is slow, and either the video is delayed with respect to the audio, or both are delayed, making conversation difficult.

We hear talk of the "bandwidth explosion." Will the situation improve soon? Personally I think not, even though there is a great deal of unused fiber on long-distance routes. High-speed home connections are available in the form of DSL and cable modems. These certainly offer an improved experience, but mainly because there are so few of them. There is little evidence that the switching infrastructure supporting these connections is growing at anything like the necessary pace. More and more Web sites are offering content that requires more bandwidth. I see an explosion of bandwidth demand and creeping infrastructure bandwidth growth.

The spread of multicasting technology will certainly help with applications such as "live" broadcast. Today, almost all Internet streaming relies on unicast operation—the server must deliver a separate bitstream to each individual user, addressed appropriately, even if thousands of users are watching a live event. When more multicast technology is available, servers will be able to deliver a single bitstream to which multiple users may subscribe.

Standards for Streaming Media

The first standard to achieve widespread adoption was H.261, an ITU Standard approved in 1990. It is intended for use on ISDN circuits and uses multiples of 64 kbits/s. It supports the common image format (CIF) at 352 × 276 pixels and the quarter-size QCIF at 176 × 144 pixels. It uses a combination of motion compensation, DCT, and VLC, and can code at rates up to 30 frames/s. Unlike MPEG, H.261 and other streaming media standards permit frame dropping as a means to control bit rate. It is very common in video conferencing systems to see the image freeze when some rapid motion, such as a pan, occurs.

H.263 followed in 1996. Initially this standard was intended for very low bit rates only (up to 64 kbits/s), but this restriction was removed and the standard is likely to replace H.261. It employs improved coding algorithms and supports QCIF and SQCIF (approximately half-QCIF). As an option, implementations can support higher resolutions up to 16 CIF at 1408 × 1152 pixels.

More recently, the H.323 series of standards provides higher quality over corporate intranets and to institutions with second-generation Internet connections. Operation is typically in the hundreds of kilobits/s range. The higher quality offered by these standards and bandwidths is particularly attractive to academic and medical conferencing.

In the world of personal computers there is intense rivalry among three companies, Apple, RealNetworks, and Microsoft. The original format, and still very popular, is *Quicktime* by Apple. This format supports a wide range of bit rates and compression schemes—nowadays almost anything can be made into a Quicktime movie. On the CD included with this book you will find commercial stock footage at full resolution, and low bit rate preview versions of the same clips; both are supplied as Quicktime movies. There is a free version of the Quicktime player, but Apple does not seem to offer any free version of the encoder.

RealNetworks range of products including *RealAudio* and *RealVideo* have probably shown the fastest growth in recent years as many web sites have added audio and video elements. As mentioned previously, we now expect to hear audio samples from CDs before we buy them online. The RealNetworks solutions have proven very attractive at the low bit rates available on today's Internet. RealNetworks offer a free basic player, and a trial version of the encoder is also available for download without charge. More sophisticated players, and fully functional encoders, are available for a price.

The most recent major contender is Microsoft Corporation. It has mounted a major effort to redefine streaming media with the *Advanced Streaming Format* (ASF), offering end-to-end solutions. Microsoft offers a complete range of tools including players, encoders, servers, and supplementary tools at its *Windows Media* site. Microsoft's new Media Player, available at this site and included with the latest operating systems, supports a wide range of coding formats (except Real), and the latest encoder supports MPEG-4 (Version 1 at the time of writing). The CD includes sample Windows Media files, at audio/video bit rates ranging from 28 kbits/s to 2Mbits/s. Microsoft's determination to become a substantial player in the field of streaming media may be deduced from the fact that (at the time of writing) *all* of these tools are available without charge.

Some of the tools mentioned above will be found on the CD-ROM included with this book; others may be obtained from the respective websites. In all cases, it would be wise to check the websites for the latest versions of these products.

Microsoft Corporation has mounted a major effort to redefine streaming media with *Advanced Streaming Format* (ASF) offering end-to-end solutions. Microsoft offers a full range of free tools, including encoders, players, and servers, plus a wealth of information at its Web site. Microsoft's new Media Player, available at this site, and included with the latest operating systems, supports a vast range of coding formats (except Real), and the latest encoder supports MPEG-4 (Version 1 at the time of writing).

Some of these tools are on the CD-ROM included with this book, but check out the Web sites for latest versions.

CHAPTER **19**

Closing Thoughts

Introduction

This chapter is the dumping ground for all the bits and pieces that would not fit elsewhere. For example, I take some time here to look at the issue of concatenated compression systems—a subject of great importance since compression is used in acquisition, in editing, and in delivery. There is a discussion of mezzanine, or "in between," compression systems that may offer a partial solution to some of these issues.

The use of compression systems for television distribution presents many new challenges. Common functions, such as the insertion of commercials at local stations or at cable head ends, may depend on the ability to switch compressed bit streams. We examine some of the difficulties inherent in this process.

Finally, we take a brief look at some future directions—those that are fairly close, and some that perhaps require "suspension of disbelief."

Fractal Compression

Fractal compression requires enormous computing power for encoding, but a relatively simple decoder. Fractal compression takes a block of pixels, and attempts to find another block that can be transformed into the current block. Rotation, reflection, scaling, gain, offset, and the like are all allowed. If large parts of the image can be reduced to a recursive relationship, these parts of the image may be represented by a relatively small number of parameters of a *contractive mapping*.

Some workers claim spectacular results with this method, but as yet it has found little general acceptance. Perhaps the most significant claim of fractal compression is that it produces a file that may be rendered back at any resolution, lower or higher than the original. Of course, no additional detail is created by enlarging the image, but the fractal approach ensures that typical artifacts of enlargement, such as pixel blocking, do not appear.

The CD-ROM with this book includes two fractal-based applications for compressing static images. I encourage you to experiment with these and compare the results with those obtained from DCT- and wavelet-based applications (also on the CD). The fractal programs are not commercial products and do not have the same degree of "polish" as the for-sale applications, but are suitable for experiment. It is interesting to note that the only fractal-based commercial program I know of has been withdrawn (though the shareware version is still available at some Internet locations).

Statistical Multiplexing

In Chapter 9 we saw that temporal compression necessarily results in variable bit rate at the output of the compression encoder. The efficiency gained by predictive encoding in P- and B-frames means that they will be smaller than I-frames; conversely, for an I-frame to be a good reference for generating P- and B-frames, it must be as error free as possible—in other words, as big as possible. Chapter 9 described the type of rate control mechanism and the buffers necessary to interface an MPEG encoder to a fixed-rate transmission circuit.

The above discussion addresses only a part of the problem, in that it assumes constant complexity of the video being encoded. In practice, video varies a great deal in complexity. A "talking-head" shot is easier to encode than a scene with moving objects. A static graphic is easier still; at the other extreme, a zooming shot of complex motion in a basketball game, with a crowd behind, is about as difficult as video can get. It is fairly obvious that there will be both short- and long-term variations. A sport program is intrinsically more "bit hungry" than a talk show, but within each program there will be some parts that are much more complex than others. Unfortunately, even the short-term variations are likely to have time scales of tens of seconds or minutes, so increased buffering is not a practical solution.

The problem is a real one; at a fixed bit rate greater complexity means lower quality, lesser complexity means higher quality. Unfortunately, the human psychovisual system is much more sensitive to variations in quality than it is to absolute quality, and such variations are likely to be visible and annoying.

In a system with a single program stream and fixed bit rate there is little to be done, except to try to design the system so that the lowest quality received (most-complex video) is sufficiently good that improvements in quality are not readily apparent. If this is achieved, there is a choice—the quality can be allowed to vary, or the quality can be held down to the acceptable level and the surplus bits used for some other purpose or just "stuffed."

Some applications allow variation in bit rate. A movie on DVD (previously digital video disc or digital versatile disc, now just DVD) uses an average bit rate of about 5 Mbits/s, but the DVD system is capable of a peak bit rate of some 10 Mbits/s. This allows the compression device (or the human compressionist) considerable leeway in the allocation of bits to scenes of different complexity.

Another solution is available when a fixed-rate MPEG transport bit stream carries a number of program streams (as described in Chapter 10), as used by direct broadcast satellite systems. Statistical multiplexing is a technique based on the assumption that not all of the program streams will carry complex video at the same time. Within the fixed-rate bit stream, at any instant more bits can be allocated to the program streams carrying complex video, and fewer to those carrying simple video, stills, and other data. Three approaches are available to control the distribution of bits:

- The bit allocation of a program may be determined by its type; a football game will need more bits than a sitcom, and a movie coded at 24 frames per second will probably need even less bits than the sitcom. In a simple system this determination can be used to divide the bit allocations between the various program streams. In a more complex system with dynamic control of bit allocation, program type may be used for the initial estimate of bit rate requirements for each program stream.

- Using conventional rate control techniques, information from the rate controller of each compression encoder can be considered by a statistical multiplex controller. This controller will provide a modifier control back to each rate controller so that encoders operating at a low quantization scale factor will be instructed to quantize more aggressively, thereby freeing up additional bits for encoders that are being forced to use a high scale factor. In this way, the statistical multiplex controller can balance out the bit allocations so that all program streams are using a similar scale factor and providing a similar level of quality.

- The most sophisticated controllers may use *look-ahead* techniques to measure the complexity of upcoming frames (as may be done by the encoder itself). This information may be used to further "tune" the rate controllers. For example, if several streams have complex material arriving shortly, it may be appropriate to increase all quantization scale factors to reduce buffer fullness and provide for a peak in bit requirements.

The above techniques work best if the transport stream is carrying a considerable number of program streams, and if there is a good mix of program types. Obviously, statistical multiplexing is complex and expensive; however, the payoffs are considerable. Use of this technique not only averages out the impairment level of each program stream, but for a given quality level it allows more program streams within a channel of given capacity. In fact, the gain is even higher in the commercial world; if the quality is not varying appreciably, a lower quality can be tolerated. (Expressed simply, the human psychovisual system, and viewers' psychology, will accept a given quality level if it is not shown how much better it could be!) These factors make it possible to substantially increase the number of program streams of acceptable quality in a given channel. In commercial terms, each satellite transponder can carry more programs. Given that each transponder in place represents an investment of tens of millions of dollars, the advantage is compelling.

Concatenated Compression Systems

I have discussed at length the sensitivity of compression systems to artifacts and noise on the signal to be encoded. These can cause substantial losses in efficiency. A compression system adds artifacts and quantization noise to an image sequence, so what happens when we feed one compression system with the output of another? This is far from an idle question. We are seeing increasing use of compression systems for acquisition and in studio operations, and we are about to see the large-scale introduction of heavily compressed digital transmission systems. There would seem to be ample reason for concern.

Experimental results suggest that compression systems are very tolerant of their own artifacts. In most of the compression systems in use today, a quality loss is apparent on the first pass, but if the signal is not changed in any way, subsequent passes through an identical compression system add almost no loss.

In one sense we can see that this is intuitively reasonable. A compression system discards information, particularly information that should have little effect on the perceived image. A second pass through an identical system attempts to throw away the same information, but it is no longer there. The second system should be able to code the remaining information just as happily as the first.

Jim Wilkinson of Sony first explained this to me as a theoretical model. Most of the elements in the compression encode and decode process have inverse elements, and most combinations are lossless. For example, variable-length encoding and decoding is lossless, so it need not be considered as part of the system. The combination of DCT and inverse DCT (IDTC) is lossless with sufficient arithmetic precision. This pair of elements contributes round-off errors, but little else.

The lossy process is, of course, quantization. There is a block called dequantization, but it serves only to return the coefficients to the correct reconstruction values; the original information is irretrievably lost.

However, once this step has been performed, we can again look at a lossless system. Assuming there is no signal processing between the output of the decoder and the input to the next encoder, we can follow the values. Quantized coefficients are passed via variable-length encode/decode (lossless), then to the inverse DCT, thus producing a set of sample levels. If DCT/IDCT is nearly lossless, so is IDTC followed by DCT. After the DCT process of the next encoder, we will have the same coefficient set that was produced by the dequantizer. Each of these levels should be right in the center of the decision range for the new quantizer, so that it should produce the same result. Figure 19-1 illustrates.

Figure 19-1
Lossless groupings in concatenated compression systems.

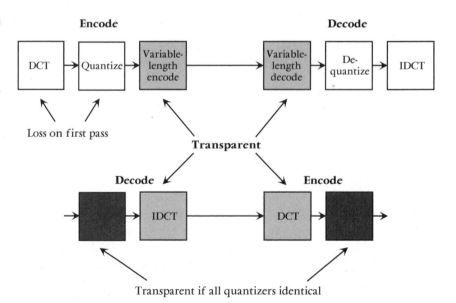

Considered in this way, the explanation of the observed behavior is simple. Quantization causes the loss observed on the first pass. On subsequent passes the same quantization results should occur every time.

The only issue is the DCT/IDCT match, which is never perfect. Arithmetic precision errors and round-off errors contribute the very gradual deterioration in performance on subsequent generations.

It will be seen that the quantizer is an essential element of this process. The lossless situation exists only if the quantizers are identical. Successive coding by two encoders that use different quantizers will likely result in quite rapid deterioration.

Researchers at Sarnoff Corporation have shown that compression systems can be very sensitive to the artifacts produced by other compression systems. In one experiment, two "light" compression systems intended for studio operations with high-definition television were compared. One was based on MPEG, the other on wavelets. The signals, both of which were free from visible artifacts, were fed to an ATSC transmission encoder that uses MPEG compression.

The transmission encoder performed normally when fed with the output of the MPEG system, but displayed a much higher level of artifacts when fed from the wavelet system.

This should be cause for concern. Already, television plants are full of compressed nonlinear edit systems and tape formats that use compression. We are about to see widespread deployment of MPEG-based delivery systems, yet we see the rapid growth of DV compression, which uses a quantization strategy quite different from MPEG. None of this will cause an overnight disaster, but we may be eroding our quality headroom without even realizing it.

Two parts of a European project are of great interest in this area. In one, developers are achieving essentially lossless recoding, even with a change of bit rate. All the coding decisions (motion vectors, etc.) of the original compression are extracted from the decoder and sent along with the video. (There is a proposal to code this data by using the least significant bit of one of the color difference signals.) The same coding decisions can then be used by the second encoder—a big step toward retaining quality.

The other work specifically concerns changing bit rates—perhaps from a distribution signal to a broadcast delivery signal. The encoder uses a special form of quantizer that recognizes the previous quantization decisions and attempts to minimize the combined effect of the two quantizers. Demonstrated results are spectacular.

Audio compression systems are not immune to the problems of concatenation. As with video, a compression system designed for final delivery to the point of use should be designed so that the effects of quantization are undetectable. If two identical systems are concatenated then, just as with video, the decisions in the second system should match those of the first and result in little or no additional impairment, *if the signal has not changed.* If the signal does change, the likely result is that a coarsely quantized signal is requantized differently, and it is very likely that the impairment will be unacceptable.

If nonidentical compression systems are concatenated, rapid deterioration is possible, again just like video. The artifacts of one system are likely to disrupt the efficient operation of the other. An interesting and very important example is the interaction of Dolby Surround Pro Logic (DSPL) with Dolby Digital (DD), if wrongly used, as described in the previous chapter.

Switching MPEG

MPEG was conceived for a relatively simple scenario in which information is encoded, sent to the decoder, and decoded, without any intervening steps other than transmission and/or storage. It became evident that there were situations where it would be desirable to switch from one program stream to another, or to a different part of the same stream. An obvious example in the context of MPEG's early objectives is the interactive movie—a story that switches to alternative routes and endings according to actions of the viewer.

Because of these considerations, a number of provisions were made within MPEG for switching or *splicing* of the bit stream. However, these techniques were really designed for the type of situation just described, where there is prior determination of the switch-out points and of the potential switch-in points. There is no mechanism in MPEG for a generalized switch from any point in one bit stream to any point in another.

Some closer thought makes it clear why this cannot be. At the lowest level, a variable-length—encoded bit stream is meaningful only if decoded from the beginning. Arbitrary switching provides no way to know what the data is, or what it means. If we tried to switch bit streams, like those shown in Chapter 7, we could (and likely would) end up interpreting extra bits as Huffman codes; what should have been motion vectors would be wrongly decoded and, perhaps, used as if the values were DCT coefficients.

We noted earlier that one of the important characteristics of the MPEG slice is that it is encoded without any outside references. All predictive encodings are reset at the beginning of a slice, so that we can begin decoding unambiguously at the start of any slice. Actually this demonstrates that we have not yet defined the problem. We could start decoding the new stream at any splice, but it is not productive to begin decoding a picture in the middle. Resynchronization at a slice boundary is useful for recovery from data errors, but not for switching between programs.

Perhaps we can use the picture boundaries for switching, just as we use vertical interval switching of baseband video today. For I-frame—only encoding we can do this, subject to certain conditions. I-frame—only streams, like motion JPEG streams, can be switched with reasonable ease. Even at this point there are complications. If we permit the same number of bits for each frame (the simplest case), a frame is 33.4 ms long in the nominal 60-Hz regions (such as North America) and 40 ms long in 50-Hz areas (such as Europe and Australasia). Neither lines up with the 32 ms used by a frame of AC-3 audio. Typically this means leaving a small gap in the audio whenever a switch is made, and this complication exists however or wherever we switch.

With other GOP structures the situation is more complex. Suppose we switch to the new bit stream, and the first picture is a P-frame. Some macroblocks will likely be intracoded, and these will decode successfully. Others, however, will be predictively encoded—the bit stream will carry a motion vector identifying a similar macroblock in the previous I-frame and a set of DCT-coded information representing the changes that should be made to that macroblock. The assumption made by the encoder is, of course, that the I-frame is present in storage at the decoder. Well, there is an I-frame in the decoder store, but it belongs to the previous video sequence—the one before the switch (Figure 19-2)! If other conditions are correct the decoder will use it, but it will produce nonsense.

If the frame following the switch is a B-frame, the situation is only slightly worse. The B-frame may use both forward and backward prediction, on the assumption that both of the surrounding pair of anchor frames (*I, I* or *I, P* or *P, P*) are in the decoder memory. Probably there are two frames in memory, and they may even be a pair of adjacent anchor frames, but both will belong to the previous video stream (Figure 19-3).

Figure 19-2
Incorrect P-frame
reference after a
splice.

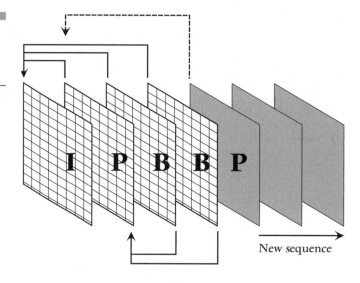

Figure 19-2
Incorrect P-frame
reference after a
splice.

Figure 19-3
Incorrect references
from a B-frame.

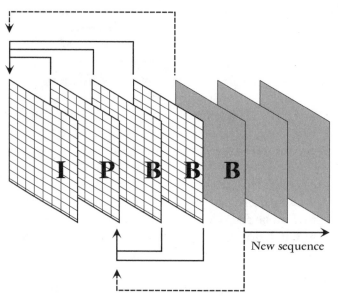

It appears that we may be able to switch MPEG bit streams only at GOP boundaries, even though this may not provide sufficient resolution for some applications. We will look at some applications in a moment, but there is one more major issue awaiting us, even at the GOP boundary.

In Chapter 9 we looked at MPEG rate control and discussed the concept of the VBV buffer model. Although not based on real buffer designs, VBV provides a means of specifying decoder buffer performance and a means of checking that a bit stream is decodable by a standard decoder. The basic conditions are that VBV should never overflow or underflow.

VBV is filled at a nominally constant rate by the transmission channel. It is emptied at regular (frame) intervals, but by varying amounts depending on whether the frame being extracted is an I-, P-, or B-frame, and on the complexity of the particular frame. VBV represents an important degree of freedom in the battle for constant quality, and a good decoder will use the full gamut of the buffer to help achieve this. At times the buffer will be close to overflow, at other times close to underflow.

It all works because the encoder tracks VBV fullness at all times; except, of course, when we switch bit streams! The encoder for the second stream is managing VBV for hypothetical decoders that have been receiving the bit stream *since the beginning of the sequence*. If any (or all) of the decoders have been receiving a different sequence, there is no way for the encoder to "know" that the switch has occurred, or that the decoder VBV is in an unknown and unknowable state. We might switch to an almost full VBV at a time when the second stream encoder has calculated that VBV should be almost empty; overflow is inevitable. Similarly, switching to a near-empty VBV when the encoder is tracking VBV as close to full will result in underflow.

What happens when VBV overflows or underflows is undefined by MPEG, and depends on the design of a particular decoder. Most likely (and this is probably the least disturbing solution) the picture will freeze until the decoder finds that it again has valid data waiting to be decoded. Unfortunately, the complexity of the process makes it uncertain how quickly this point will be reached. It is also quite possible that the decoder will start decoding again, but with a VBV state that still differs from that assumed by the encoder. In this case another overflow or underflow is likely, but perhaps some considerable time after the switch.

As a final twist, statistical multiplexing might be the issue that makes everything else look easy. It is difficult to imagine how to switch into one program stream—that is, to replace all packets associated with that one program in the multiplexed bit stream—when the number of bits allocated to that program changes dynamically under the control of some device that may be on the other side of the country. A working scenario would have to include rules about bit rate floors, at least at commercial insertion times, and bit rate ceilings for any material to be inserted at the switch.

Clearly, switching MPEG bit streams is a complex and uncertain process. In fact, the preceding discussion presents only a gross simplification of the problem. Almost certainly the general problem for two unconstrained bit streams is insoluble. An MPEG expert would suggest that MPEG is not designed for such treatment, so "Don't do that!" Unfortunately, in the world of television broadcasting there are a number of areas where switching of MPEG streams is the only practical solution to an operational requirement. Let's look at some applications and see how they might be handled.

MPEG Applications

MPEG is used by recording devices and it would be useful to edit segments together without decoding. On disk recorders particularly, the ability to perform edits merely by using the random access capabilities of the disk is a major convenience. At the time of writing, almost all disk systems use motion JPEG, or some other intracoding scheme where splicing is not a problem. However, there are strong incentives to use MPEG for increased recording time and/or higher image quality. MPEG's increase in efficiency comes mainly from temporal compression, so intracoding is not an option that satisfies these requirements.

The ATSC *Digital Television Standard* uses MPEG-2 encoding for video and provides for (nominally) one high-definition program or several multiplexed program streams of standard definition. Normally about 18 Mbits/s is available for video in the transmission signal. Initially it was assumed that this signal would be distributed by networks, and that television stations would switch to local commercials by switching the MPEG bit stream. The lack of production values offered to the station, and the difficulties of the switching, have contributed to the unpopularity of this scheme, and most networks are now proposing a different approach (see below).

As we move down the transmission food chain, however, there is a continued need for switching, and although neither is attractive, eventually there is no option but to decode and re-encode, or to switch the transmission signal. The most obvious example is the cable head-end. As cable systems extend the digital signal to the home, economics demand that the head-end operate in pass-through mode, without decoding the video and audio. Yet insertion of local commercials at cable head-ends is now a thriving business, and this requirement will certainly not disappear.

Most of these problems are as yet unsolved, and it is clear that there is no universal solution, but potential solutions for various applications are on the horizon.

Some Solutions

There are significant developments in the field of re-encoding, based on reusing the original coding decisions. These developments will make a decode/re-encode approach viable in many areas by drastically reducing the losses normally associated with process.

Another approach requires partial decoding of the signal to change frame types to preserve the integrity of motion compensation. In particular, B-frames may be modified so that vectors are used in only one direction. At the International Broadcasting Convention (IBC) in 1977, the European Atlantic project demonstrated switching of this type.

Conventional wisdom might suggest that the first frame of the new sequence (after a switch) should be an I-frame, requiring no references from previous frames. However, B-frames may be used if they employ only backward prediction—that is, prediction based on an anchor frame later in time. (But remember, the I-frame still must be transmitted first, even though it is for later presentation.) The advantage is that these B-frames can, at the expense of image quality, be squeezed to a very low number of bits if required, and it may be necessary to reduce the number of bits at the beginning of the second sequence to correct the VBV fullness.

Taking bits from the B-frames has two advantages. First, the B-frames are a part of the sequence that can be deprived of bits without affecting the quality of the I-frame or introducing errors that will propagate. Second, a scene change provides a substantial degree of temporal masking in the human visual system. If the low-quality B-frames occur immediately after the switch, it is most unlikely that the quality loss will be noticed.

Certainly the IBC demonstration supported this view. An example switch between two sequences used two unidirectional B-frames immediately after the cut. Examining the result frame by frame revealed that these B-frames were of very poor quality, but at normal play speed the same sequence was quite acceptable, with no apparent quality loss.

It is not certain what impact such developments will have on decisions on signal switching at cable head-ends and similar locations. If a switch can be made by decoding and re-encoding (switching at video) with no significant loss of quality, this mechanism may be attractive for many applications. It may be that, by combining the techniques of reusing coding decisions and modifying frame types, the motion detector could be eliminated from the re-encoder. This would eliminate the most significant cost element and make this approach financially attractive.

In the meantime, most workers assume that some method of direct switching of MPEG bit streams will be required for some applications. A working group of SMPTE, under the chairmanship of the author, investigated possible techniques for some two years. The real work was performed by an ad hoc group chaired by Katie Cornog of Avid, with members from many different industries. Eventually the committee produced a standard for splicing MPEG bitstreams, SMPTE 312M, published in 1999.

The standard specifies how MPEG transport streams may be constructed to permit splicing. Splice points, in-points, and out-points are inserted at the desired points in the bit stream, and messages may be sent in the bit stream to advise splicing devices of upcoming splice points.

The key to necessary VBV buffer management is to relate buffer fullness to a delay. As the "input" to VBV is a constant rate bit stream, each value of buffer fullness corresponds to a certain delay: the time it would take for the buffer to reach fullness from empty. This is known as the *splice decoding delay*. A *seamless* splice is defined as one where there is no discernible artifact when the spliced bit stream is played by a standard decoder. To achieve a seamless splice, both bit streams are constrained to have a splice decode delay equal to a defined nominal value at the splice point. (There are many other constraints, but this is the essential factor for VBV management.)

To provide for cases where a seamless splice cannot be achieved, the proposed standard also defines a *nonseamless* splice where VBV may undergo a controlled underflow. A well-designed decoder will play a seamless splice with minimal disturbance, usually a freeze of the video for a few frames.

Since the publication of the SMPTE Standard, it has become increasingly apparent that for most applications, seamless splicing is not practical. The bitstreams require considerable preparation to include splice points and to manage the buffers in the vicinity. It appears that other techniques, making use of mezzanine compression, transparent decode/encode, and/or GOP modification will provide practical solutions for studio applications.

For lower-budget applications, such a local commercial insertion in cable systems, the cable industry has developed economical approaches that will likely be incorporated into a future version of the SMPTE Standard.

Mezzanine Compression Systems

It is clear from the preceding sections that switching compressed signals is not simple. Furthermore, production effects other than cuts are needed frequently at intermediate points. For example, a local television station must switch local commercials to a network feed, but also must key station identification and alert messages over the video and insert voiceovers in the audio. Some stations perform more sophisticated production effects such as squeezing the network image as the credits roll to allow a "promo" for the next program. At present none of these effects can be performed on the compressed signal—to produce spatial effects we need the signal in the spatial domain, not the frequency domain. It is possible that solutions may be found for the simple cases, but in general it is safe to assume that production effects require a baseband (uncompressed) signal, whether in video or audio.

This presents a problem. Although we have seen that identical compression systems concatenate almost losslessly, this is not true if the signal is changed between decoding and re-encoding. In the case of MPEG delivery systems, it is fair to assume that insufficient headroom exists to be able to decode, perform effects, and re-encode without substantial degradation.

The issue arises at many points in a television broadcast chain, in contribution and distribution channels, and in the TV station when recording devices are encountered. Most standard-definition digital video recorders use some form of compression, and it is highly probably that compressed recording will dominate the world of high definition. How do we design a viable system that includes compression at so many points, and where there is a need for production effects at many different points in the chain?

The concept of *mezzanine*, or in-between, compression is evolving to meet these needs for broadcast systems based on the ATSC Digital Television Standard. (To quote a colleague, a mezzanine is the floor in a hotel you can never find, and where your meeting is!)

Requirements for a mezzanine system cover quite a large range. At one extreme, the mezzanine signal must be carried over a single satellite transponder (around 45 Mbits/s maximum) as part of the network distribution, and must have sufficient quality overhead to permit decoding, production effects, and re-encoding to the ATSC emission standard (approximately 18 Mbits/s for video).

At the other extreme, compression used for recording, and possibly in-plant distribution, must be sufficiently robust to permit a significant number of decode/re-encode operations (perhaps 6 to 10) without noticeable degradation. It is also required that this signal be capable of simple switching and editing, so intra-only coding should be used, preferably with a fixed number of bits per frame. Color encoding should be at least 4:2:2 for studio effects.

From the discussion of concatenation, we can draw some conclusions about the required characteristics of the mezzanine systems. Working backward from the emission system defined, we must see the same algorithm (DCT), the same DCT block placement, and the same quantization strategy.

Obviously, there is a scheme that meets these requirements—MPEG-2 itself—but as published there is no MPEG profile/level combination that meets all these needs. Because of the urgent need for a working document, SMPTE is in the process of standardizing a new operating point, the 4:2:2 profile at high level (4:2:2@HL) permitting bit rates up to 300 Mbits/s. Table 19-1 shows some examples of how this might be used. (SDTI is *serial data transmission interface*—a 1998 SMPTE standard for transmission of packetized data over serial digital links in the studio.)

TABLE 19-1

Possible scenarios for the use of MPEG for pre-transmission encoding of high-definition TV signals.

Application	Possible bit rates	Possible GOP structures	Possible circuits
Contribution (low delay)	155 Mbits/s	I only, or IP or IB	OC-3 fiber
Contribution (other)	60-80 Mbits/s	Long GOP if delay unimportant, otherwise balance between quality and delay requirements	8-PSK satellite
In-plant	200-270 Mbits/s	I only	Recorders and SDTI
Distribution	40-45 Mbits/s	Same as emission standard	QPSK satellite or DS-3

Similar issues arise with audio distribution, but there are some different twists. Like any compression system, AC-3 (Dolby Digital) can be decoded and re-encoded, but this is not desirable when compression has been performed on the final delivery rate. However, production requirements are similar—a station needs to perform voiceovers, and these cannot be done with audio in a compressed form. It is possible to use the same compression scheme at a higher data rate for distribution, and this is certainly one possibility. However, once the need to deviate from the transmission standard is seen, it may be profitable to explore further. In this case requirements in the studio suggest an alternative approach that could also be used for distribution.

Two problems arise in the studio. One is the number of channels required. The six channels could be carried by three AES/EBU circuits, but most recording devices will not handle this many channels. Also, it is difficult if not impossible to carry three AES/EBU circuits in perfect synchronism through all the operations and equipment of a television studio. It is probable that the signals will slip by at least 1 bit. Unfortunately 1 bit represents a phase change of some 35° at 10 kHz, and this is a significant error in a true surround-sound system. The other problem is that there is no convenient mechanism to carry the metadata (discussed in Chapter 17) that is required for the AC-3 encoder.

Dolby Laboratories has designed a new compression scheme to satisfy these requirements. The new system, Dolby-E, uses an algorithm based on that used for AC-3 and specifically designed to be benign when used ahead of AC-3. It will compress 5.1 channels into the payload of a single AES/EBU circuit, together with the metadata needed for the AC-3 encoder. Some versions of Dolby-E provide additional channels within the same bitstream. This can be particularly useful when a program needs 5.1 audio for digital transmission and stereo or DSPL for analog transmission.

Dolby-E also addresses an issue common to all other digital audio systems—the fact that the length of the frame or access unit does not relate easily to the length of the video frame. Dolby-E has versions corresponding to all the major video formats, and in each case the length of the access unit matches that of the video frame. This removes many of the problems traditionally associated with switching and editing digital audio and video.

This audio mezzanine approach has many advantages. The bit rate is acceptable for distribution, yet high enough to permit a useful number of decode/reencode operations; stations need not triple the layers of audio routing, videotape recorders will be able to record the signal, and the metadata will be delivered as part of the package.

A Glimpse into the Future

Even though MPEG-2 is just going into widespread service, great things are being planned for the future. The MPEG-4 standard is approved and applications are appearing.

MPEG-7 is well under way and, in conjunction with new technology, promises to greatly increase the real value of multimedia archives, MPEG-21 is intended to provide a complete structure to manage digital assets.

We saw that MPEG-4 can transmit a model of a face, and then animate that face to match dialog. Other researchers are working on the implementation of *agents*—models that will not only accurately mimic the delivery of a speech, but will do so in whatever language is requested.

That's one of the interesting things about compression and its allied technologies. The progress over the last few years has been dramatic, yet the more we know, the more we realize that we have only scratched the surface. I find the subject fascinating—I just hope that I have managed to convey some of my fascination with this book.

GLOSSARY

This glossary is based upon that included in the ATSC Digital Television Standard, A53, and permission to reproduce it here is gratefully acknowledged. The ATSC glossary, and some of my additions, include terms that are not strictly applicable to video compression and will not be found elsewhere in this book. Others refer to advanced topics beyond the scope of this book. All of these terms are, however, relevant to the field of digital television. Some of the definitions are those from the MPEG standards; in these cases I have retained the formal syntax (e.g., group_start_code).

8-PSK—A variant of QPSK used for satellite links to provide greater data capacity under low-noise conditions.

8 VSB—Vestigial sideband modulation with 8 discrete amplitude levels.

16 VSB—Vestigial sideband modulation with 16 discrete amplitude levels.

ACATS—Advisory Committee on Advanced Television Service.

AC-3—The audio compression scheme invented by Dolby Laboratories and specified for the ATSC *Digital Television Standard*. In the world of consumer equipment it is called *Dolby Digital.*

access unit—A coded representation of a presentation unit. In the case of audio, an access unit is the coded representation of an audio frame. In the case of video, an access unit includes all the coded data for a picture and any stuffing that follows it, up to, but not including, the start of the next access unit. If a picture is not preceded by a group_start_code or a sequence_header_code, the access unit begins with a picture_start_code. If a picture is preceded by a group_start_code and/or a sequence_header_code, the access unit begins with the first byte of the first of these start codes. If it is the last picture preceding a

317

sequence_end_code in the bit stream, all bytes between the last byte of the coded picture and the sequence_end_code (including the sequence_end_code) belong to the access unit.

A/D—Analog-to-digital converter.

AES—Audio Engineering Society.

anchor frame—A video frame used for prediction. I-frames and P-frames are generally used as anchor frames, but B-frames are never anchor frames.

ANSI—American National Standards Institute.

asynchronous transfer mode (ATM)—A digital signal protocol for efficient transport of both constant-rate and bursty information in broadband digital networks. The ATM digital stream consists of fixed-length packets called *cells*, each containing 53 8-bit bytes—a 5-byte header and a 48-byte information payload.

ATEL—Advanced Television Evaluation Laboratory.

ATM—*See* asynchronous transfer mode.

ATSC—Advanced Television Systems Committee, the organization charged with standards documentation for the digital television system for the United States. ATSC is now an international organization with members from many countries and is working to facilitate the smooth introduction of digital television and to promote the adoption of the ATSC system in other countries.

ATTC—Advanced Television Test Center.

ATV—The U. S. advanced television system.

bidirectional pictures, B-pictures, B-frames—Pictures that use both future and past pictures as a reference. This technique is termed *bidirectional prediction*. B-pictures provide the most compression. B-pictures do not propagate coding errors, because they are never used as a reference.

bit rate—The rate at which the compressed bit stream is delivered from the channel to the input of a decoder.

bits/s—Bits per second.

block—A block is an 8×8 array of pixel values or DCT coefficients representing luminance or chrominance information.

bps—Bits per second.

byte-aligned—A bit in a coded bit stream is byte-aligned if its position is a multiple of 8 bits from the first bit in the stream.

CDTV—*See* conventional-definition television.

channel—A digital medium that stores or transports a digital television stream.

coded orthogonal frequency-division multiplex—A modified form of OFDM. A modulation scheme used for digital transmission that is employed by the European DVB system. It uses a very large number of carriers (hundreds or thousands), each carrying data at a very low rate. The system is relatively insensitive to Doppler frequency shifts and can use multipath signals constructively. It is, therefore, particularly suited for mobile reception and for single-frequency networks.

coded representation—A data element as represented in its encoded form.

COFDM—*See* coded orthogonal frequency division multiplex.

compression—Reduction in the number of bits used to represent an item of data.

conventional-definition television (CDTV)—This term is used to signify the *analog* NTSC television system as defined in ITU-R Recommendation 470. *See also* standard-definition television and ITU-R Recommendation 1125.

CRC—Cyclic redundancy check to verify the correctness of the data

D-frame—Frame coded according to an MPEG-1 mode, which uses DC coefficients only.

data element—An item of data as represented before encoding and after decoding.

DCT—*See* discrete cosine transform.

decoded stream—The decoded reconstruction of a compressed bit stream.

decoder—An embodiment of a decoding process.

decoding (process)—The process defined in the Digital Television Standard that reads an input coded bit stream and outputs decoded pictures or audio samples.

decoding time stamp (DTS)—A field that may be present in a PES packet header that indicates the time that an access unit is decoded in the system target decoder.

Digital Video Broadcasting (DVB)—A system developed in Europe for digital television transmission, originally for standard definition only, although high-definition modes have been added to the specification. DVB defines a complete system for terrestrial, satellite, and cable transmission. Like the ATSC system, DVB uses MPEG-2 compression for video, but it uses MPEG audio compression and COFDM modulation for terrestrial transmission.

discrete cosine transform (DCT)—A mathematical transform that can be perfectly undone and which is useful in image compression.

Dolby Digital—*See* AC-3.

DTS—*See* decoding time stamp.

DVB—*See* Digital Video Broadcasting.

DVCR—Digital video cassette recorder.

ECM—*See* entitlement control message.

editing—A process by which one or more compressed bit streams are manipulated to produce a new compressed bit stream. Conforming edited bit streams are understood to meet the requirements defined in the Digital Television Standard.

EIA—Electronic Industries Association.

elementary stream (ES)—A generic term for one of the coded video, coded audio, or other coded bit streams. One elementary stream is carried in a sequence of PES packets with one and only one stream_id.

elementary stream clock reference (ESCR)—A time stamp in the PES from which decoders of PES may derive timing.

EMM—*See* entitlement management message.

encoder—An embodiment of an encoding process.

encoding (process)—A process that reads a stream of input pictures or audio samples and produces a valid coded bit stream as defined in the Digital Television Standard.

entitlement control message (ECM)—Private conditional access information that specifies control words and possibly other stream-specific, scrambling, and/or control parameters.

entitlement management message (EMM)—Private conditional access information that specifies the authorization level or the services of specific decoders. They may be addressed to single decoders or groups of decoders.

entropy coding—Variable-length lossless coding of the digital representation of a signal to reduce redundancy.

entry point—A point in a coded bit stream after which a decoder can become properly initialized and commence syntactically correct decoding. The first transmitted picture after an entry point is either an I-picture or a P-picture. If the first transmitted picture is not an I-picture, the decoder may produce one or more pictures during acquisition.

ES—*See* elementary stream.

ESCR—*See* elementary stream clock reference.

event—A collection of elementary streams with a common time base, an associated start time, and an associated end time.

Federal Communications Commission (FCC)—The government agency responsible for (among other things) the regulation of spectrum utilization in the United States, and the body that licenses radio and television broadcast stations.

field—For an interlaced video signal, a field is the assembly of alternate lines of a frame. Therefore, an interlaced frame is composed of two fields, a top field and a bottom field.

forbidden—This term, when used in clauses defining the coded bit stream, indicates that the value shall never be used. This is usually to avoid emulation of start codes.

FPLL—Frequency- and phase-locked loop.

frame—A frame contains lines of spatial information of a video signal. For progressive video, these lines contain samples starting from one time instant and continuing through successive lines to the bottom of the frame. For interlaced video a frame consists of two fields, a top field and a bottom field. One of these fields will commence one field later than the other.

group of pictures (GOP)—One or more pictures in sequence.

high-definition television (HDTV)—High-definition television has a resolution of approximately twice that of standard definition television in both the horizontal (H) and vertical (V) dimensions and a picture aspect ratio (HxV) of 16:9. ITU-R Recommendation 1125 further defines "HDTV quality" as the delivery of a television picture that is subjectively identical with the interlaced HDTV studio standard.

high level—A range of allowed picture parameters defined by the MPEG-2 video coding specification, which corresponds to high-definition television.

Huffman coding—A type of source coding that uses codes of different lengths to represent symbols that have unequal likelihood of occurrence.

IEC—International Electrotechnical Commission.

intracoded pictures, I-pictures, I-frames—Pictures that are coded by using information present only in the picture itself and without depending on information from other pictures. I-pictures provide a mechanism for random access into the compressed video data. I-pictures employ transform coding of the pixel blocks and provide only moderate compression.

ISO—International Organization for Standardization.

ITU—International Telecommunication Union.

JEC—Joint Engineering Committee of EIA and NCTA.

JPEG—Joint Photographic Experts' Group.

layer—One of the levels in the data hierarchy of the video and system specification.

level—A range of allowed picture parameters and combinations of picture parameters.

macroblock—The fundamental unit within MPEG for motion prediction. A macroblock includes all data relating to a picture area defined by 16×16 luminance pixels. In the advanced television system a macroblock consists of four blocks of luminance and one each C_B and C_R block.

main level—A range of allowed picture parameters defined by the MPEG-2 video coding specification with maximum resolution equivalent to ITU-R Recommendation 601.

main profile—A subset of the syntax of the MPEG-2 video coding specification that is expected to be supported over a large range of applications.

motion vector—A pair of numbers that represent the vertical and horizontal displacement of a region of a reference picture for prediction.

MP@HL—Main Profile at High Level.

MP@ML—Main Profile at Main Level.

MP3—An abbreviation for MPEG audio, Level III. A popular format for compressing music files for distribution over the Internet.

MPEG—Refers to standards developed by the ISO/IEC JTC1/SC29 WG11, Moving Picture Experts Group. MPEG may also refer to the group.

MPEG-1—Refers to ISO/IEC standards 11172-1 (Systems), 11172-2 (Video), 11172-3 (Audio), 11172-4 (Compliance Testing), and 11172-5 (Technical Report).

MPEG-2—Refers to ISO/IEC standards 13818-1 (Systems), 13818-2 (Video), 13818-3 (Audio), and 13818-4 (Compliance).

NCTA—National Cable Television Association.

NTSC—National Television Systems Committee. The committee that defined 525-line monochrome television in the 1940s and the U.S. standard-definition color television system in the 1950s.

OFDM—Orthogonal frequency division multiplex. *See also* COFDM.

pack—A pack header followed by zero or more packets; it is a layer in the system coding syntax.

packet—A header followed by a number of contiguous bytes from an elementary data stream; it is a layer in the system coding syntax.

packet data—Contiguous bytes of data from an elementary data stream present in the packet.

packet identifier (PID)—A unique integer value used to associate elementary streams of a program in a single- or multi-program transport stream.

padding—A method to adjust the average length of an audio frame in time to the duration of the corresponding PCM samples by continuously adding a slot to the audio frame.

payload—The bytes that follow the header byte in a packet. For example, the payload of a transport stream packet includes the PES_packet_header and its PES_packet_data_bytes or pointer_field and PSI sections, or private data. A PES_packet_payload, however, consists only of PES_packet_data_bytes. The transport stream packet header and adaptation fields are not payload.

PCM—Pulse-code modulation.

PCR—*See* program clock reference.

pel—*See* pixel.

PES—Packetized elementary stream.

PES packet—The data structure used to carry elementary stream data. It consists of a packet header followed by PES packet payload.

PES packet header—The leading fields in a PES packet up to but not including the PES_packet_data_byte fields where the stream is not a padding stream. In the case of a padding stream, the PES packet header is defined as the leading fields in a PES packet up to but not including the padding_byte fields.

PES stream—PES packets, all of whose payloads consist of data from a single elementary stream, and all of which have the same stream_id.

picture—Source, coded, or reconstructed image data. A source or reconstructed picture consists of three rectangular matrices representing the luminance and two chrominance signals.

PID—*See* packet identifier.

pixel—Picture element, or pel. A pixel is a digital sample of the color intensity values of a picture at a single point.

predicted pictures, P-pictures, P-frames—Pictures that are coded with respect to the nearest *previous* I- or P-picture. This technique is termed *forward prediction*. P-pictures provide more compression than I-pictures and serve as a reference for future P-pictures or B-pictures. P-pictures can propagate coding errors when P-pictures (or B-pictures) are predicted from prior P-pictures where the prediction is flawed.

presentation time stamp (PTS)—A field that may be present in a PES packet header that indicates the time that a presentation unit is presented in the system target decoder.

presentation unit (PU)—A decoded audio access unit or a decoded picture.

profile—A defined subset of the syntax specified in the MPEG-2 video coding specification.

program—A collection of program elements. Program elements may be elementary streams. Program elements need not have any defined time base; those that do have a common time base and are intended for synchronized presentation.

program clock reference (PCR)—A time stamp in the transport stream from which decoder timing is derived.

program element—A generic term for one of the elementary streams or other data streams that may be included in the program.

program-specific information (PSI)—Normative data that is necessary for the demultiplexing of transport streams and the successful regeneration of programs.

PSI—*See* program specific information.

PTS—*See* presentation time stamp.

PU—*See* presentation unit.

QAM: *See* quadrature amplitude modulation.

QPSK—Quadrature phase-shift keying, a modulation scheme for digital transmission particularly suitable for satellite links.

quadrature amplitude modulation (QAM)—A modulation scheme for digital transmission that uses both amplitude and phase information. Favored by the cable industry for digital television distribution.

quantizer—A processing step that intentionally reduces the precision of DCT coefficients.

random access—The process of beginning to read and decode the coded bit stream at an arbitrary point.

SCR—*See* system clock reference.

scrambling—The alteration of the characteristics of a video, audio, or coded data stream to prevent unauthorized reception of the information in a clear form. This alteration is a specified process under the control of a conditional access system.

SDTV—*See* standard-definition television.

slice—A series of consecutive macroblocks.

SMPTE—Society of Motion Picture and Television Engineers.

source stream—A single, nonmultiplexed stream of samples before compression coding.

splicing—Concatenation of, or switching between, two different streams of compressed data.

sprite—A static video object, usually a background, that is larger than the presentation area. Once the sprite has been transmitted, a degree of camera movement may be represented without transmission of a new background.

standard-definition television (SDTV)—Used to signify a digital television system in which the quality is approximately equivalent to that of NTSC. Also called standard digital television. *See also* conventional-definition television.

start codes—32-bit codes embedded in the coded bit stream that are unique. They are used for several purposes including identifying some of the layers in the coding syntax. Start codes consist of a 24-bit prefix (0×000001) and an 8-bit stream_id.

STD—*See* system target decoder.

STD input buffer—A first-in, first-out buffer at the input of a system target decoder for storage of compressed data from elementary streams before decoding.

still picture—A coded still picture consists of a video sequence containing exactly one coded picture, which is intracoded. This picture has an associated PTS, and the presentation time of succeeding pictures, if any, is later than that of the still picture by at least two picture periods.

streaming, streaming media—A technique for transmitting digital content over networks and links so that the content may be presented as it is received without waiting for the end of the transmission.

system clock reference (SCR)—A time stamp in the program stream from which decoder timing is derived.

system header—A data structure that carries information summarizing the system characteristics of the Digital Television Standard multiplexed bit stream.

system target decoder (STD)—A hypothetical reference model of a decoding process used to describe the semantics of the Digital Television Standard multiplexed bit stream.

time stamp—The time of a specific action, such as the arrival of a byte or the presentation of a presentation unit.

TOV—Threshold of visibility.

transport stream packet header—The leading fields in a transport stream packet up to and including the continuity_counter field.

variable bit rate—Operation where the bit rate varies with time during the decoding of a compressed bit stream.

video buffering verifier (VBV)—A hypothetical decoder that is conceptually connected to the output of an encoder. Its purpose is to provide a constraint on the variability of the data rate that an encoder can produce.

video sequence—A video sequence is represented by a sequence header, one or more groups of pictures, and an end_of_sequence code in the data stream.

BIBLIOGRAPHY

Abramson, Norman. *Information Theory and Coding*. McGraw-Hill: New York, 1963.

Bartlett, Eugene R. *Cable Television Handbook*. McGraw-Hill: New York, 2000.

Benson, K. Blair, and Fink, Donald G. HDTV: *Advanced Television for the 1990s*. McGraw-Hill: New York, 1991.

Benson, K. Blair, revised by Whitaker, Jerry. *Television Engineering Handbook*. McGraw-Hill: New York, 1992.

Benson, K. Blair, and Whitaker, Jerry. *Television and Audio Handbook for Technicians and Engineers*. McGraw-Hill: New York, 1990.

Clarke, Roger C. *Digital Compression of Still Images and Video*. Academic Press: San Diego, 1995.

De Bruin, Ronald and Smits, Jan. *Digital Video Broadcasting: Technology, Standards and Regulations*. Artech House: Norwood, Mass., 1999.

Habibi, A. "Comparison of N-th Order DPCM Encoder With Linear Transformations and Block Quantization Techniques." *IEEE Transactions on Communications*, COM-12(6): 948—956, 1971.

Hendee, William R. and Wells, Peter N.T. *The Perception of Visual Information*, 2d ed. Springer-Verlag: New York, 1997.

Hubbard, Barbara Burke. *The World According to Wavelets*. 2d ed. A.K. Peters, Ltd.: Wellesley, Mass., 1996.

Inglis, Andrew F. *Video Engineering*, 2d ed. McGraw-Hill: New York, 1996.

Jack, Keith. *Video Demystified*, 2d ed. HighText Publications: San Diego, 1996.

Lim, Jae S. *Two Dimensional Signal and Image Processing*. Prentice Hall: Englewood Cliffs, N.J., 1990.

Lohscheller, H. "Subjectively Adapted Image Communication System," *IEEE Transactions on Communications*, COM-32(12): 1316—22, December 1984.

Microsoft Corporation. *Inside Windows Media*. Macmillan USA: Indianapolis, 1999.

Mitchell, Joan L., Pennebaker, William B., Fogg, Chad E., and LeGall, Didier J. *MPEG Video Compression Standard*. Chapman & Hall: New York, 1996.

Netravali, Arun N., and Haskell, Barry G. *Digital Pictures*. Plenum Press: New York, 1988.

Pennebaker, William B., and Mitchell, Joan L. *JPEG Digital Image Compression Standard*. Van Nostrand Reinhold: New York, 1993.

Poynton, Charles A. *A Technical Introduction to Digital Video*. John Wiley & Sons: New York, 1996 (out of print; see superceding edition below).

Poynton, Charles. *Digital Video and HDTV: Pixels, Pictures, and Perception*. John Wiley & Sons: New York, 2001.

Pratt, William K. *Digital Image Processing*. John Wiley & Sons: New York, 1978.

Rabbani, Majid, and Jones, Paul W. *Digital Image Compression Techniques*. SPIE Optical Engineering Press: Bellingham, Wash., 1991.

Reza, F. M. *An Introduction to Information Theory*. McGraw-Hill: New York, 1961.

Robin, Michael and Poulin, Michel. *Digital Television Fundamentals: Design and Installation of Video and Audio Systems*, 2d ed. McGraw-Hill: New York, 2000.

Rogowitz, Bernice E., and Allebach, Jan P., eds. *Human Vision and Electronic Imaging*, SPIE Proceedings, Bellingham, Wash., 1996.

Schreiber, William F. *Fundamentals of Electronic Imaging Systems*, 2d ed. Springer-Verlag: New York, 1991.

Shannon, C. E. "Prediction and Entropy of Printed English," *Bell System Technical Journal*, 30(1): 50—64, January 1951.

Taylor, Jim. *DVD Demystified*. McGraw-Hill: New York, 1998.

Turin, William. *Digital Transmission Systems: Performance Analysis and Modeling*. McGraw-Hill: New York, 1999.

Watkinson, John. *The Engineer's Guide to Motion Compensation*. Snell & Wilcox Ltd.: Petersfield, U.K., 1994.

Watkinson, John. *Compression in Video and Audio*. Focal Press: Oxford, U.K., 1995.

Watkinson, John. *The Engineer's Guide to Compression.* Snell & Wilcox Ltd.: Petersfield, U.K., 1996.

Watson, Andrew B., ed. *Digital Images and Human Vision.* MIT Press: Cambridge, Mass., 1993.

Weiss, S. Merrill. *Issues in Advanced Television Technology.* Focal Press: Boston, 1996.

Welch, Terry A. "A Technique for High Performance Data Compression," *IEEE Computer,* 17(6): 8—19, June 1984.

Whitaker, Jerry C., *DTV: The Revolution in Digital Video,* McGraw-Hill: New York, 1999.

Whitaker, Jerry C. and Benson, K. Blair, eds. *Standard Handbook of Video and Television Engineering,* revised ed. McGraw-Hill: New York, 2000.

Whitaker, Jerry, ed., *Electronics Handbook.* CRC Press: Boca Raton, Fla., 1996.

INTERNET RESOURCES

There is a wealth of information available from the Internet on the subjects of video and audio compression. I have listed just a few of the key sites; those that I expect to remain stable. The best way to research the Internet is to use the links from these sites; generally these are kept up to date. To save typing, these links may be found in the top directory of the CD-ROM.

http://www.symes.tv   This is where I will post updates to the book, errata, and links to new or newly discovered interesting Web sites. It also provides a mechanism for feedback on the book and related topics. If you find errors, or have suggestions for improvements, please use the Web site and let me know.

http://www.mpeg.org/index.html/   This is not the "official" MPEG site, but it is without doubt the richest source of information on almost anything connected with compression. There is even a site search engine! The site is published by MpegTV LLC, a San Francisco-based privately owned company founded in 1997, and maintained by Tristan Savatier, a long-time contributor to MPEG. Most, if not all, of the following sites may be accessed from the approximately 2,000 links indexed in mpeg.org.

http://drogo.cselt.stet.it/mpeg/   This is the official MPEG site. A visit to this site is particularly useful for those interested in the ongoing work of MPEG. There are some excellent tutorials on completed and current work.

http://www.m4if.org/   The industry forum for MPEG-4. News, information, and links to companies with MPEG-4 products.

http://www.jpeg.org/   A great deal of information on JPEG, including the continuing work of the committee.

http://www.dvpa.com/links/   Links to many sites covering DV compression.

http://www.real.com/   The home of Real Video and Real Audio; free players and trial versions of authoring programs, many sources of streaming media.

http://www.microsoft.com/windows/windowsmedia/en/ default.asp   Everything you ever wanted to know about Windows Media, and all the tools you need to do it. Free (at the time of writing) players, authoring software, and servers.

http://www.vide.gatech.edu/cookbook2.0/   Excellent source of information on video conferencing.

http://www.internz.com/compression-pointers.html   Many links to sites addressing every aspect of compression technology.

http://www.atsc.org/   The home of the Advanced Television Systems Committee and the *ATSC Digital Television Standard* (A/53). Other documents available from this site include *Digital Audio Compression (AC-3) Standard* (A/52), and *Guide to the Use of the ATSC Digital Television Standard* (A/54) and an excellent tutorial on compression and digital transmission.

http://www.dvb.org/   The European Digital Video Broadcasting organization maintains extensive information on the DVB standards.

http://www.unik.no/~robert/hifi/dvd/   Extensive information on *DVD*; what was *digital video disk* or digital versatile disk is now just *DVD*.

http://www.davic.org/   The Digital Audio Visual Council is an association of some 200 companies seeking interoperability in the multimedia communications world.

http://www.smpte.org/   The Society of Motion Picture and Television Engineers is an international standards organization. SMPTE publishes the studio standards that form the input to standardized compression systems and many standards relating to compression.

http://www.iec.ch http://www.iso.ch http://www.itu.ch   Official sites of the standards organizations, including information on the purchase of standards documents.

http://www.dolby.com/   Many general interest and technical papers on audio systems and audio compression.

http://www.inforamp.net/~poynton/   Charles Poynton's home page with a fascinating collection of articles on digital signal processing, color science, and other aspects of vision and television.

http://www.math.auth.gr/~axonis/studies/audio.htm A site created by Panos Stokas of the Department of Mathematics at Aristotle University in Thessaloniki, Greece. It describes a long-term exercise to find the best algorithm for audio compression, one that combines high quality at low bit rates with processing speed and portability.

http://www.sv.philips.com/newtech/ Good information on many audio and video related technologies.

ABOUT THE CD-ROM

The CD-ROM included with this book includes a number of software applications for the compression of still images, video, and audio, together with some sample source material. All applications are intended for the Microsoft Windows® operating system. All of the content is provided for educational and experimental purposes, and no license is granted to use any material or application for any commercial purpose.

The contents of the CD are listed in a readme file in the top-level directory. On Windows 2000 and Windows ME systems, this file should be opened automatically in your browser when you insert the CD. If not, open this file manually and proceed from there. All of the content is described in linked html files, so you can investigate the CD from your browser. If you do not have Version 5.5 of Internet Explorer installed, you might want to start by installing this program from the CD.

Because the content has been supplied by multiple organizations, you will find considerable variety in the organization and presentation. Generally the content from each source will be in separate directories, together with any particular applicable instructions, copyright notices, and the like.

Some of the applications are freeware (but still subject to licensing agreements); others are demonstration versions that may have limited functionality, or that may be used only for a specified time after installation. In most cases the application or the accompanying files will contain explanations, links to the company that owns the application, and instructions on how to purchase unrestricted versions (and some of the applications are very inexpensive).

I believe it is helpful to gather all this content together in a way that allows easy exploration, but I strongly suggest that if you find an application useful you should contact the manufacturer to obtain the latest version. Please be aware of the licensing conditions for each product, and respect the rights of those who own the content. In particular, please do not redistribute any of the content of the CD unless the license for that content explicitly permits redistribution.

INDEX

ABOUT THE AUTHOR

Peter Symes is Manager of Advanced Technology at Grass Valley Group. He is an SMPTE Fellow, a Senior Member of the IEEE, and the author of *Video Compression* (McGraw-Hill). At the time of this writing, he served as the Engineering Director, Television for SMPTE.